THE SERIES OF "CHINA'S MARITIME DEVELOPMENT"

CHINA'S MARINE CONSERVATION AND DEVELOPMENT

AUTHORS: LIU YAN, QIU JUN, ZHENG MIAOZHUANG, ZHU XUAN

 China Intercontinental Press

图书在版编目（CIP）数据

美丽海洋：中国的中国的海洋生态保护与资源开发：英文 / 刘岩等编著；刘美英译 . -- 北京：五洲传播出版社，2014.9

（中国海洋系列）

ISBN 978-7-5085-2948-6

Ⅰ.①美… Ⅱ.①刘… ②刘… Ⅲ.①海洋生态学－中国－英文②海洋资源－资源开发－中国－英文 Ⅳ.① Q178.53 ② P74

中国版本图书馆 CIP 数据核字 (2014) 第 256811 号

"中国海洋"丛书

策　　划：付　平
出 版 人：荆孝敏
主　　编：张海文　高之国　贾　宇

美丽海洋——中国的海洋生态保护与资源开发（英文）

编　　著：刘　岩　丘　君　郑苗壮　朱　璇
特 约 编 辑：何北剑
责 任 编 辑：黄金敏　姜　超
翻　　译：刘美英
审　　校：郑惠贞
图 片 提 供：国家海洋局海洋发展战略研究所　中国新闻图片网
　　　　　　东方 IC　FOTOE　CFP
装 帧 设 计：丰饶文化传播有限责任公司
出 版 发 行：五洲传播出版社
社　　址：北京市北三环中路 31 号生产力大楼 B 座 7 层
电　　话：0086-10-82007837（发行部）
邮　　编：100088
网　　址：http://www.cicc.org.cn　http://www.thatsbooks.com
印　　刷：北京市艺辉印刷有限公司
开　　本：710mm×1000mm　1/16
字　　数：149 千字
图　　数：120 幅
印　　张：14.75
印　　数：1－5000
版　　次：2014 年 12 月第 1 版第 1 次印刷
定　　价：86.00 元

Preface

On April 12th, 1961, the world's first manned spacecraft "Oriental" was launched in the Soviet Union. Gagarin described in his historic space flight the scene that human beings have never seen, "The horizon presents an exceptionally beautiful view. A light blue halo encircles the Earth and is mingled with the black sky. In the sky, bright stars are shining with well-defined boundaries." After the 108-minute epochal flight which circled the Earth once, Gagarin returned to Earth.

In December 1968, the U.S. spacecraft "Apollo 8" started its trip to the Moon. Bill Anders, astronaut in the spacecraft, captured the most classic picture in the American aerospace history–the blue Earth rises up from the horizon of the gray moon. This marked the first time for human beings to see the panorama of the Earth.

Anousheh Ansari, an American businesswoman and also the world's first female space tourist, recalled, "The sheer beauty of it (the Earth) just brought tears to my eyes."

How we see the world determines how we treat the world. And how we see the world in turn determines how we treat the oceans.

The blue part of the Earth that we see from space is the ocean, which accounts for 71% of the total area (360 million square kilometers) of the Earth's surface. The oceans contain over 1.35 billion cubic kilometers of water, accounting for about 97% of the total amount of water on Earth. So far, humans have explored only 5% of the seabed, leaving the remaining 95% of the seabed unknown to the public. As the land resources for human survival become increasingly scarce, the oceans with relatively infinite resources have long been the hope for future survival of humanity.

However, when we see the Earth and the oceans from the perspective of the whole universe, and see the oceans from the perspective of human existence, we will find that the oceans have long been scarce and fragile resources for human beings.

Height from which people see things determines their view, and their vision determines their attitude.

In 1990, the 45th UN General Assembly rendered a decision, urging all countries to take marine development and use as a national development strategy. In 1992, the United Nations Conference on Environment and Development passed the Agenda 21 which pointed out: Oceans are a fundamental part of the world's life support system, and also a valuable asset to help achieve sustainable development. On November 16th, 1994, the United Nations Convention on the Law of the Sea (UNCLOS) came into effect, marking the establishment of a modern system of international maritime laws, and laying a legal foundation for the sustainable development of global marine resources and environments.

With development of human beings and advance of the world, China, as a developing country, must have a new vision and attitude towards the oceans and their development, while making leaps and bounds in its economic and social development.

With a coastline stretching about 18,000 kilometers long, mainland China has about 7,300 islands, each covering an area of more than 500 square meters, and governs a sea area of about 3 million square kilometers. China seas cover three climatic zones of warm temperate zone, subtropical zone and subtropical zone, with the coastal waters being featured by various types of marine ecosystems, such as mangroves, coral reefs, coastal salt marshes, seagrass beds, sea islands, bays, estuaries, and upwellings, as well as abundant marine resources including fisheries, energies, ports and sceneries. Thousands of rivers run into the seas. According to incomplete statistics, there are over 1,500 rivers in six major drainage basins of the Liao River, the Haihe River, the Yellow River, the Huaihe River, the Yangtze River, and the Pearl

River running into the seas. Rich natural environments have created colorful marine environments, making them a beautiful part of China.

China's rich and diverse marine resources and marine environments have made huge ecological services and resources support the for the economic and social development in its coastal areas and even the entire China, acting as a valuable asset to the sustained and healthy economic development in China. Currently, the main marine resources available for development and utilization include marine biological resources, mineral resources, seawater resources, marine renewable energies and marine spatial resources.

With the development and use of marine resources, over 10 marine industries have been formed. The added value created by these industries accounts for about 10% of China's GDP. Meanwhile, the marine resources development activities have also provided over 33 million jobs.

On the one hand, the oceans have provided tremendous support for national economic and social development; but on the other, the rising enthusiasm for marine development and the rapid development of marine economy have also brought enormous pressures on the oceans especially on the ecological environment in the offshore, which has seen sustained deterioration in recent years. At the same time, under the impacts of both human activities and climate changes, disasters occur frequently to Chinese seas, causing huge economic losses. The sustainable development of Chinese seas faces greater challenges.

The Chinese government attaches great importance to environmental protection and sustainable development of Chinese seas, and actively participates in the process of global sustainable development. It has formulated the China's Agenda 21-White Paper on China's Population, Environment and Development in the 21st Century, which specified the sustainable development strategies to be implemented for the future development of China. China is a country with both a vast land and a long coastline. China's social and economic development will increasingly depend on the oceans. Therefore, the

China Agenda 21 took "sustainable development and protection of marine resources" as one of the important action plans.

In the 21st century, the Chinese government pays more attention to the development of the marine industry, giving it a priority in the national economic and social development. With the continuous improvement of sustainable development policies, the sustainability of the marine industry also increases steadily.

In China, the environmental protection of marine ecosystem and the scientific development of marine resources under the premise of protection, have received unprecedented praise and respect from all social circles. Although it is a long process, as this process becomes a forward direction of the world, China's progress in this area will not reverse, but will even be accelerated to approach the target. The situation where "ecological environment" heroes are respected by people is gradually formed. The understanding of building beautiful oceans being the common responsibility of mankind is becoming a reality.

Contents

Prosperity Beyond the Land
——China's Marine Ecosystems and Resources — 11

Diverse Coastal Ecosystems — 12
Abundant Marine Resources — 21
Valuable Offshore Space — 40

"Vitamin" to China's 1.3 billion Population
——China's Utilization of Marine Resources — 51

No. 1 in Mariculture Production in the World — 54
Complete Category of Marine Product Processing Industry — 56
Four Marine Biotechnology and Drug Research Centers — 57
"Going Global" and "Inviting In" for the Exploitation of Oil and Gas Resources — 58
Integration and Industrialization of Seawater Utilization — 61
Utilization of Marine Renewable Energy — 66

Positive Energy in the Face of Natural Disasters
——Prevention and Reduction of Marine Disasters in China — 73

Marine Disasters — 74
Prevention of Marine Natural Disasters — 92

The Azure Ocean —China's Marine Environment Conditions in Coastal Waters 99

Offshore Water Quality —— 102

Typical Ecosystem Health —— 103

Red Tides and Green Tides —— 104

"Dead Zones" in Offshore Waters —— 113

Oil Spills —— 115

Land-sourced Pollutants —— 118

Ecological Impact of Land Reclamation from the Sea —— 121

Protect the Source of Life —Ecological and Environmental Protection of China Seas 125

The Establishment of Marine Environmental Protection System —— 128

Implacement of Marine Functional Zoning —— 130

Establishment of System of Island Protection —— 132

Marine Environmental Protection Mechanism to Explore Sea Linkage —— 134

Maintaining Marine Fishing Resources in Multimodal Means —— 136

Ecological Restoration for Seas and Islands —— 139

Speeding Up the Construction of Marine Protected Areas —— 142

Constitution of a Comprehensive Monitoring System in Marine Environment —— 148

To Achieve Normalization of Marine Environmental
   Protection Law Enforcement ——————————— 150

Best Practice of Marine Nature Reserve ——————— 155

Future of Marine Development —— China's Sustainable Marine Development Strategies 165

Progress of Sustainable Marine Development ————— 166

China's Capacity for Sustainable Marine Development ——— 183

Sustainable Development Path of China'Seas ————— 201

Contrusting the Ecosystem-based Comprehensive
   Marine Management ————————————— 218

Conclusion 228

Schedule 229

CHINA'S MARINE CONSERVATION AND DEVELOPMENT

PROSPERITY BEYOND THE LAND
——CHINA'S MARINE ECOSYSTEMS AND RESOURCES

Diverse Coastal Ecosystems

China seas cover three climatic zones of warm temperate zone, subtropical zone and subtropical zone. There are a number of rivers in China running into the seas and covering a wide drainage area. The diverse natural environments have given birth to the various types of marine ecosystems. China has a wide range of marine biological resources, accounting for 10% of the global marine species, and has one of the five coastal waters with the world's richest marine biodiversity.

More than 1,500 Rivers Running into the Sea

On its land of 9.6 million square kilometers, China has a coastline of more than 18,000 kilometers, where more than 1,500 large and small rivers flow into the seas, with their runoff volume accounting for 69.8% of the total runoff volume of all rivers in China. Among them, the Yangtze River, Yellow River, Qiantang River and Pearl River are featured by a wide watershed area and a large river runoff.

With a total length of 6,300 kilometers and a basin area of over 1.8 million square kilometers, namely about one-fifth of the total land area of China, the Yangtze River is China's largest river and also the world's third longest river. It originates from the southwest side of Geladandong Snow Mountain—the peak of the Tanggula Mountain on the Qinghai-Tibet Plateau—and its mainstream flows through 10 provinces, municipalities and autonomous regions in China into the East China Sea. The average annual

CHINA'S MARINE ECOSYSTEMS AND RESOURCES

The picturesque scenery along the Yellow River in Ji'nan, at dusk.

amount of water flowing from the Yangtze River into the seas is up to 1 trillion cubic meters, ranking third in the world.

As the second longest river in China, the Yellow River is famous for its muddy water. It originates from the Yueguzonglie Basin at the northern foot of the Bayan Har Mountain, traverses 7 provinces in China and empties into the Bohai Sea in Kenli County of Shandong Province. The Yellow River is 5,464 kilometers long, with a basin area of 752,400 square kilometers. It carries an average annual amount of more than 1 billion tons of sediments into the Bohai Sea, creating the modern Yellow River Delta.

The third largest river in China is the Pearl River. The main tributaries of the Pearl River extend about 11,000 kilometers long and cover a total basin area of 452,600 square kilometers.

The Qiantang River is the largest river in Zhejiang Province in China, with a total length of 605 kilometers and a basin area of over 48,800 square kilometers. It originates in the southwest of Xiuning County of Anhui Province, with its mainstream flowing through Anhui and Zhejiang provinces, and finally into the sea after going across the Hangzhou Bay.

There are a large number of rivers and streams flowing into China seas, not only bringing in a lot of fresh water, but also carrying and dissolving abundant substances due to the wide basin area in mainland China. They have exerted significant impacts on the natural environment in China seas.

17 Estuaries

Although China has more than 1,500 small and big rivers flowing into the sea in its coastal zones, there are only 17 important estuaries. They are the Yellow River estuary, the Yangtze River estuary, the Pearl River estuary, the Tumen River estuary, the Yalu River estuary, the Liao River estuary, the Luan River estuary, the Hai River estuary, the Guan River estuary, the Qiantang River estuary, the Jiaojiang River estuary, the Ou River estuary, the Min River (Fujian, China) estuary, the Jiulong River estuary, the Han River estuary, the Nanliu River estuary and the Beilun River estuary.

> In oceanography, the wide lower part of a river where it flows into the sea is called an estuary, also a river estuary or a river mouth. It is where a river is connected with the sea. Seeing from above the sea, an estuary is a semi-enclosed coastal water body, connecting an open ocean, into which one or several large rivers flow. On the two sides of an estuary, lots of sediments are usually deposited due to tides when rivers flow into the sea. When salty seawater is mixed with and diluted by freshwater discharged from inland areas, an area rich in aquatic products will be formed.

Estuaries are one of the most important marine ecosystems, and also the area with the highest productivity, rich biodiversity, and maximum intensity of development and utilization in oceans.

Rivers of various sizes flow into the sea to form a large number of estuary ecosystems, which are featured by rich biological diversity and particularity. In terms of particularity, take aquatic organisms as an example. As one of the important wetlands in China and also an area with the most

Satellite Imagery of the Yangtze River Estuary

dense distribution and the most complete species of white dolphins within Chinese territory, the Pearl River estuary possesses the only psephuyrus gladii and eriocheir sinensis in the world. In addition, it has many major migration routes or short stay areas for anadromous species, and acts as a spawning and breeding place for lots of important economic animals.

World's Largest Reed Marsh Wetland

If you drive southwest from Panjin City in China's northeast Liaoning Province for nearly an hour, you will reach the reed protection levee which spreads across the seaside. Standing on the reed protection levee, you can see Liaohe River Delta Wetland, the so called world's largest vegetation type—the well-preserved reed swamp.

Coastal salt marshes, located at estuaries between land and sea, usually grow a variety of halophytes, herbaceous plants, and intertidal benthos. The reed community is where halophytes and herbaceous plants are most widespread in coastal salt marshes in China.

The Liaohe River Delta is a reed marsh located at the intersection of the Liaohe River and Daliaohe River in the Bohai Bay, with an area of 1.2

Red-crowned Cranes in the Panjin Wetland

million mu. As an important community of coastal wetlands in northern China, this coastal wetland is featured by biodiversity. It not only grows reeds and red suaeda glauca, but also serves as an important habitat of birds. More than 250 kinds of birds inhabit here, including red-crowned cranes, endangered species of larus saundersi, whooper swans, and ciconia boyciana. According to the research by experts, the Liaohe River Delta is the southernmost line where the red-crowned crane breeds and the world's largest habitat for the larus saundersi. Meanwhile, it is also an important node on the migration routes of migratory birds in East Asia.

Mangrove Ecosystems

> Mangrove ecosystems are a distinctive ecosystem at estuaries. They are a continuum of ecological functions, constituted by mangrove plants grown on tropical coastal mudflats and their surrounding environment.
>
> In a mangrove ecosystem, the main plant species are mangrove, rhizophora mucronata, ceriops tagal, kandelia candel, bruguiear gymnorrhiza, and bruguiera sexangula. They have breathing roots or prop roots; they breed inside the fruit: When the fruit is still on the tree, the seeds are sprouting into seedlings, and after getting ripe, they get rid of and jump from the big tree with small branches onto the beach, flow with the seawater, and settle down in the right place.

Mangrove ecosystems are distributed in some coastal beach areas of Fujian, Taiwan, Guangdong, and Guangxi in China. Among them, the most abundant resources of mangroves are in Guangxi, where one third of China's mangrove grows. China's mangrove is the most representative on the west coast of the Pacific Ocean, whether in species or in distribution range.

Mangroves play a significant role in the prevention of coastal disasters and are also the habitats for animals in coastal areas and on beaches. They can protect against wind and dissipate waves, induce siltation and protect the beaches, consolidate the banks and reinforce levees, and purify water and air to bring ecological benefits.

Mangrove ecosystems are one of the few ecosystems with the most diverse species in the world so far. For example, Shankou mangrove reserve area in Guangxi Province of China has 111 species of benthic macrofauna, 104 species of birds, and 133 species of insects. Moreover, it has 159 species and varieties of algae, of which four species are new in China.

The Nature Reserve of Mangroves at Dongzhai Harbor in Hainan is located in the northeastern part of Hainan. Dongzhai Harbor stretches for 50 kilometers and covers more than 40 square kilometers. The Nature Reserve of Mangroves in Dongzhai Harbor is not only the first of its kind established in China, but also the largest in China. Mangrove covers an area of 15.78 square kilometers. In this area, Mangrove species are divided into 133 types, among which are 22 true mangrove species and 11 semi-mangroves, accounting for 90% of the mangrove species in China.

Tourists are boating in the lush mangroves in the Dongzhaigang Mangrove Nature Reserve in Haikou City, Hainan Province.

Shankou Mangrove National Nature Reserve is located on both sides of the Shatian Peninsula, Hepu County, Beihai City, Guangxi Zhuang Autonomous Region. The reserve is in the subtropical climate and stretches along the coast for 50 kilometers. Covering a total area of 7,000 square kilometers and with 15 mangrove species, the reserve is the second national mangrove nature reserve in China.

The Nature Reserve of Mangroves in Zhanjiang City of Guangdong Province is located in the southernmost part of mainland China, scattered like a ribbon on the beach of Leizhou Peninsula in the southwest of Guangdong Province. Mangrove covers an area of more than 70 square kilometers. In this area, there are 25 kinds of mangrove species, 194 bird species, 130 shellfish species and 139 fish species.

30,000 km^2 of Islands Composed by Coral Insects

The vast South China Sea is dotted with more than two hundred islands, reefs and submerged shoals. They are like gems inlaid on China's seas. But do you know what these islands are composed of? They are composed of coral insects!

From the latitude perspective, two thirds of China's seas are in tropical and subtropical zones, suitable for the growth of corals. Coral reefs are mainly distributed in Xisha and Nansha Islands and along Taiwan and Hainan coasts. According to rough estimation, the total area of coral reefs on South China Sea islands is about 30,000 square kilometers.

Coral reefs are unique animals in tropic shallow waters. They are small in size and large in quantity, and can breed quickly. They adhere to rocks and grow on them, drinking "water" and secreting calcium carbonate which forms their hard bones. The new generation sticks to and grows on the bones of the old generations. After thousands of years, the small coral insects become beaches and islands made of coral reefs in the ocean. For 30 million years, coral insects have multiplied over a hundred species in the South China Sea, forming the South China Sea islands.

Beautiful coral reef in the Sanya National Coral Reef Nature Reserve

In addition to mangroves and coral reefs, China's seaside ecosystems also include seagrass resources in coastal areas from south to north, especially in Guangdong Province, Guangxi Province, and Hainan Province. The distribution of seagrass beds is in Hainan Province are concentrated in coastal waters from Wenchang City to Sanya City in the east of City Hainan Island and from Chenmai City to Dongfang City in the west; Guangdong's seagrass meadows are mainly distributed in waters near Liusha Bay of Leizhou Peninsula, East Island of Zhanjiang City and Hailing Island of Yangjiang City; Guangxi's seagrass meadows are mainly distributed in waters near Hepu County and the Pearl Harbor.

Abundant Marine Resources

Marine resources refer to the substances and energy that have a direct relationship with sea water and seabed. As a category of natural resources, marine resources mean resources that are formed and exist in waters or oceans, including creatures living in the sea, chemical elements dissolved in seawater, energy generated by waves, tides and currents in the sea, heat stored in the sea, mineral resources hidden on the seashore and in the continental shelf and the deep seabed, as well as the pressure difference and concentration difference formed by seawater.

In a broad sense, marine resources also include all the space and facilities available to the people's production, living and entertainment.

China's marine resources are closely related to the quality of life and survival prospects of Chinese people.

Over 26,000 Species of Marine Organisms Living in China's offshore

On a Friday afternoon in April 2014, a senior editor of the www.hellosea.net was invited to deliver a speech on the ocean for students in a prestigious high school in Shanghai. At the end of the speech, she asked the students "Do you know the total number of the species of organisms living in China's seas?" After a long time had passed, no one could give her the answer. The students, showing vacant eyes, desired to know the answer. "There are over 26,000 species," she said. Hearing the number, most of the students were wide-eyed with a surprised look on their faces.

On April 29th, 2014, the Marine Life Pavilion of QingDao Underwater World under the theme "Dream of the Blue" opened and visitor were enjoying transgenic fluorescent fish, blue devil fish, liking walking in the beautiful ocean.

To be frank, the question asked by the editor is indeed difficult to answer. How much does the Earth weigh? How many oceans and continents are there on the Earth? How many countries are there in the world? For such questions, people who have had physics classes and geography lessons probably know the answers. However, few people know how many organisms there are in seas.

In recent years, a number of Chinese cities have built or are building aquariums, such as Dalian Sun Asia Ocean World, Qingdao Polar Ocean World, Shanghai Ocean Aquarium, and Xi'an Qujiang Ocean World. Even Dalian City of Liaoning Province has built a world-class and China's only venue—the Polar Museum—for polar marine animal shows and polar experience; in the museum, there are large marine mammals such as beluga whales, dolphins and sea lions, which can do wonderful performances; in the Marine Animal Museum of the Polar Museum, visitors can get up close to and play with the sea lions, feed them personally, and experience the friendly coexistence of animals and humans.

However, for nearly 1.4 billion Chinese people, the number of aquariums and polar museums is, after all, too small. So the number of people who can have direct knowledge of the oceans is also too small.

However, now it is time for Chinese people, especially Chinese pupils, to face a larger number of more specific and deeper questions about the oceans. This fresh, necessary and important knowledge can not only bring people excitement and joy, but also allow people to experience the novelty, mystery and fun of marine life, especially the dispensability of the oceans for human survival.

China has vast and beautiful oceans where there is a variety of marine life closely related to the prospect of the survival of Chinese people.

People usually cherish beautiful things as well as materials related to the survival of each person. When both are facing a crisis, people will rise up to protect them. Oceans belong to these beautiful things and materials. It is time for us to recognize, understand and protect them.

Marine Biological Resources Providing the Chinese People with 1/5 of the Total Animal Proteins

Living marine resources refer to marine resources that have life, can reproduce on their own, and can constantly update. They are composed of different types of marine animals, marine plants, fungi and deep-sea genetic resources. Many marine animals and plants are edible and are a quality food treasure of great potentials. It is estimated that oceans are providing the Chinese people with 1/5 of the animal proteins. There are still a lot of natural flora and fauna resources and genetic resources in the oceans to be developed. The coastal waters can also become an artificial sea ranch and a large-scale food production base.

In the first three quarters of 2013, the fishery economy in Liaoning

Province maintained a continuous growth momentum, with a total output value of RMB 102.81 billion, an increase of 12.4%. The value added in fishery economy and the production of aquatic products increased by 10.5% and 6.6%, reaching RMB 50.06 billion and 3.904 million tons respectively.

The harvest in the fishery industry and the growth in seafood sales volume indicate Chinese people's strong demand for seafood, and also illustrate the importance of marine products in Chinese people's lives.

There are 1,694 species of fish in coastal waters of China, including more than 250 species in the Yellow Sea and the Bohai Sea, over 600 species in the East China Sea, and more than 1,000 species in South China Sea. There are over 150 species of fish in coastal waters of China with great economic and use values, mainly include trichiurus lepturus, pasembur, squid, eel, larimichthys crocea, larimichthys polyactis, butterfly fish, silvery pomfret, mackerel, red fish, nemipterus virgatus, Gadus morrhua, sardine, sea bream, and puffer fish; molluscs with economic value include squid, cuttlefish, abalone, scallops and octopus; arthropods include prawns, freshwater shrimps, lobsters, acete chinensis, trachypenaeus curvirostris, scylla serrata, and portunid; echinoderms include sea urchin, sea cucumber and thelenota ananas; and cnidarians include jellyfish.

China's offshore fishery resources are mainly distributed in various fisheries in the Bohai Sea, the Yellow Sea, the East China Sea and the South China Sea. In the Bohai Sea area, there are lots of rivers running into the sea, providing important natural conditions for the spawning and fattening of various fishery resources, and also giving the Bohai Sea the reputation of "the cradle of fishes and shrimps". The fishery resources in the Bohai Sea mainly consist of prawns, acete chinensis, trachypenaeus curvirostris, portunid, brown shrimps and jellyfish with a short life cycle and fewer food habits. The production of such catches accounts for about 72%-75% of the total catches in the Bohai Sea; pelagic fishes mainly include mackerel, pomfret, scomberomorus niphonius, ilisha elongata, scaled sardine, mullet, and carassius auratus, while demersal and near demersal fishes mainly

On January 3rd, 2009, hundreds of fishermen returned with yellow croaker, pomfret, squid, octopus and other fresh fish in the busy coastal wharf in Ganyu, Lianyungang City, Jiangsu Province.

include trichiurus haumela, baby croaker, yellow croaker, larimichthys polyactis, johnius belengerii, ray, left-eyed flounders, bass, batterfly fish, and tonguefish.

The Yellow Sea is where various species of fishes hibernate in winters. The coastal and offshore waters of the Yellow Sea mainly produce trichiurus lepturus, larimichthys crocea, larimichthys polyactis, pomfret, codfish, ilisha elongate, left-eyed flounders, cleisthenes herzensteini, Spanish mackerel, chub mackerel, clupea pallasi, prawns, trachypenaeus curvirostris, acete chinensis, and cuttlefish.

The East China Sea is rich in fishery resources both in variety and quantity. There are five species of chub mackerel, trichiurus lepturus, larimichthys crocea, larimichthys polyactis, and navodon septentrionalis with a historic annual output of more than 100,000 tons in the East China Sea area, ranking top among the four major sea areas in terms of the number of varieties of a single high-yielding species. The catches in the East China

On July 27th, 2010, the "Sea Shell" Exhibition was held in Zhejiang Museum of Natural History in Hangzhou. More than 1,000 shell specimens with over 600 varieties, almost having all subjects of sea shells including those of from South America, West Africa, Australia, Antarctica and Asian countries were displayed.

Sea area are mainly composed of demersal and near demersal warm-temperate species, accounting for about 30% of the total catches in the sea area, followed by the pelagic fishes, accounting for about 20% of the total catches in the sea area.

The South China Sea area has high temperatures, fast plankton production and a wide variety of fish, and is an important tropical fish farm in China. The catches in the South China Sea area are mainly composed of demersal fish, accounting for over 50% of the total catches in the sea area, mainly including conger eel, saurida, surmullet, branchiostegus argentatus, and priacanthus tayenus. The production of pelagic fish can account for about 30% of the total catches of the South China Sea, of which sardinella aurita and stolephorus commersonnii account for a large proportion; decapods, crabs and cephalopods account for small proportions in the total catches of the South China Sea, about 5.5% and 2% respectively. Decapods mainly include marsupenaeus, penaeus semisulcatus, Penaeus penicillatus, and lobsters.

Shellfish and algae are also important living marine resources. There are more than 110,000 species of shellfish existing in the world and nearly 4,000 species of shellfish known in China, a considerable part of which are marine shellfish. China is the country with the largest scale and the largest number of artificial breeding of marine shellfish in the world. The major aquaculture species are scallops, oysters, and abalones. China has over 10,000 species of algae, which are divided into 10 categories, with the green algae, brown algae, red algae and blue algae mainly for medicinal use.

Deep-sea Creatures

In the depths of the ocean, there are many discovered and undiscovered hydrothermal vents, such as magmas and currents. Around these hydrothermal vents grow many biomes. They are deep-sea biological genetic resources. These biomes grow in large quantity without relying on the traditional photosynthesis. They survive and thrive by relying on their own chemical process. This unique eco-gene system has important academic value in biological research, revelation of the origin of life and expansion of the living space.

The bio-genetic resources in the deep-sea environment where humans have rarely set foot in are irreplaceable. It is estimated that there are about 450 million tons of deoxyribonucleic acid (DNA) in the 10-centimeter space at the top of the deep-sea sediments. Today when the terrestrial biological resources have been utilized to a large extent, the collection and research of deep-sea organisms and their genetic resources will provide a new way and biological materials for the development of bio-engineering technologies in biopharmaceuticals, green chemicals, water pollution treatment and oil recovery. In addition, similar to early Earth environment, the environment

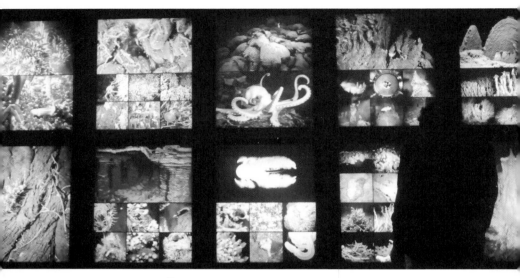

On September 29th, 2010, the "Rare Treasure in the Deep Sea" Exhibition was held in Zhejiang Science and Technology Museum. Large, medium and small deep-sea biological specimens, as well as high-resolution photographs of deep-sea creatures from the sea areas over 4,000 in depth were displayed.

in deep-sea hydrothermal vents is not only a window to observe the deep structure of the Earth, but also considered as the best place to explore the mystery and origin of life.

China's collection and research of deep-sea bio-genetic resources is still lagging behind developed countries, but China is rich in bio-genetic resources. With the increase of investments in capital and scientific research personnel, the possibility to achieve leapfrog development and make breakthrough achievements in this area is gradually becoming a reality.

"Landscapes" under the Sea

On the early morning of April 3rd, 2011, the expedition ship "Alucia" tossed violently in a squall on the South Atlantic Ocean. On the aft deck, the crew huddled together in rain slickers and gazed across the heaving sea at a yellow blur on the horizon. This yellow blur was Remus 6000 unmanned reconnaissance submarine, one of the most advanced underwater search vehicles on Earth.

In the eyes of "Remus", submarine was not all covered by the sand, but distributed with various terrains of hills, plains and valleys, with each terrain being unfathomably complex.

Mountains and central ridges under the sea are one of the most mysterious places on Earth. Two hundred and fifty million years ago when Pangea split into seven pieces of continents, the extension of landmass torn the land fissures on the Earth, forming separate canyons to be filled by water. But under the sea, volcanos were still erupting and rolling magmas, forming mountains of basalt. Today, those mountains wind across the Atlantic Ocean, through the South Pacific, and extend to the lower Indian Ocean, wrapping up the Earth like sewing a tattered football. They form the longest mountain range on the planet, totaling 50,000 miles long(1 mile equals to 1.61 kilometer). Almost no one has explored them. They are as high as the Andes, towering two miles under the sea.

The area of the ocean is three times larger than that of the land. In the ocean, there are all things that humans can see on land, including plains, valleys, mountains, sands and magmas, and even something that humans cannot see on land. It can be said that human exploration of the seabed is just at the beginning. Therefore, no one can say the mineral resources in the ocean are fewer than those on land.

Marine mineral resources include all mineral resources in coastal areas, shallows, deep seas, ocean basins, and ocean ridges and bottoms. With the current technologies, the marine mineral resources proven by people are very rich, including offshore oil and gas, gas hydrates, coastal sands, polymetallic nodules, cobalt-rich crusts and volcanogenic massive sulfide ore deposits. The total marine mineral resources have exceeded the land mineral resources. The proven resources mainly include alluvial gold, placer

A photographer is taking pictures of undersea volcanic lava cave.

platinum, diamond, placer tin, iron placer, titanium stone, zircon, ilmenite, rutile, monazite, apatite, glauconite, barite, seabed manganese nodules, and seabed oil and gas resources.

Take oil and gas resources as an example. According to the estimation by an American petroleum geologist, there are about 78 million kilometers of marine sedimentary basins worldwide with oil and gas prospects, which is generally the same with that of on land. The world's potential reserves of oil and gas less than 300 meters deep under the sea are 235.6 billion tons. The oil reserves proven in offshore seabed are 22 billion tons and the natural gas reserves 181.46 trillion cubic meters, accounting for 24% and 23% of the world reserves respectively.

China has over one million square kilometers of continental shelf with offshore depth of less than 200 meters. On the continental shelf, there are 7 sedimentary basins with oil and gas prospects, that is, the Bohai Sea, the South Yellow Sea, the East China Sea, Taiwan, Pearl River estuary, Yinggehai basin and Beibu Gulf basin, with a total area of around 700,000 square kilometers.

At present, China has exploited oil in the Bohai Sea, Beibu Gulf, Yinggehai basin and Pearl River estuary. In China's Liaodong Peninsula, Shandong Peninsula, and the coasts of Guangdong and Taiwan, there are also abundant coastal placers, mainly including gold, ilmenite, magnetite, zircon, monazite and rutile.

Oil and Natural Gas

China has over 1.3 million square kilometers of offshore continental shelf, including nearly 700,000 square kilometers of oil and gas basins. It has about 300 sedimentary basins available for exploration, of which 18 are medium and large Cenozoic sedimentary basins and 10 are large oil and gas basins: the Bohai Sea Basin, North Yellow Sea Basin, South Yellow Sea Basin, East China Sea Basin, Western Taiwan Basin, Pearl River Estuary Basin, Qiongdongnan Basin, Beibu Gulf Basin, Yinggehai Basin and

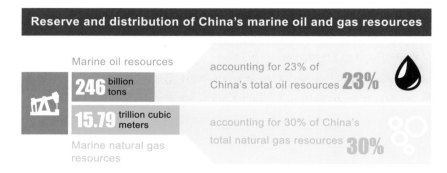

Mainly distributed in **Bohai Sea, Pearl River Estuary, southeast of Hainan, Beibu Gulf and East China Sea** 6 oil and natural gas basins

Taiwan Shoal Basin. China has more than 400 proven oil-bearing structures, of which 32 are offshore oil and gas fields including 16 in the Bohai Sea, 1 in the East China Sea and 15 in the South China Sea. In addition, China has also discovered sedimentary basins containing oil and gas in offshore areas of Okinawa Island, Western Taiwan, Northern Wan-an Bank, Western Triton Island, Northwest Palawan Island, Reed Bank, Taiping Island and James Shoal.

The quantity of marine oil resources reaches 24.6 billion tons, accounting for 23% of the total oil resources in China. The quantity of marine gas resources reaches 15.79 trillion cubic meters, accounting for 30% of the total gas resources in China. The marine oil and gas resources are mainly concentrated in the six oil and gas basins of the Bohai Sea, Pearl River estuary, Qiongdongnan, Yinggehai, Beibu Gulf and East China Sea.

At present, the amount of proven marine oil is only 3 billion tons with a proven rate of 12.3%, and the amount of proven marine gas is 1.74 trillion cubic meters with a proven rate of only 11%, far below the world's average proven rates. China has enormous potentials in exploration and development of marine resources.

Gas Hydrates

Gas hydrates refer to ice-like crystalline substances formed by gas and water under the conditions of high pressure and low temperature. Since they look like ice and can burn when meeting with fire, they are also called "combustible ice". Natural gas hydrates are distributed in deep-sea sediments or terrestrial permafrost, and are a new type of efficient energy.

In 1999, China started marine surveys of natural gas hydrates. The investigation and evaluation of natural gas hydrate resources were carried out with focuses and at different levels in four sea areas of Xisha Trough in the North Slope of the South China Sea, Shenhu, Dongsha, and Qiongdongnan, which found favorable areas for natural gas hydrates in the North Slope of the South China Sea and determined the target areas favorable for the prospects of natural gas hydrates in the North Slope of the South China Sea. In 2007, China determined 11 target areas favorable for the prospects of natural gas hydrate resources in the North Slope of

Natural gas hydrate

the South China Sea, with resources of more than 18.5 billion tons of oil equivalent, and prospective resources of 68 billion tons of oil equivalent.

On March 12th, 2013, Japan announced that it had successfully separated methane gas from natural gas hydrates contained in offshore stratums, which marked that Japan had taken a crucial step in the commercialization of natural gas hydrate exploitation. According to Japanese estimates, the potential reserves of combustible ice in waters surrounding Japan are equivalent to 100 years' natural gas consumption in Japan.

Coastal Placers

Coastal placers refer to the useful placers gathered via sorting of offshore hydrodynamics, and are featured by large scale, high grade, shallow bury, loose deposition, and easy mining and selection. The coastal placers mainly include diamond, gold, platinum, cassiterite, chromite, iron placers, zircon, ilmenite, rutile, and monazite.

Due to carriage of a large number of minerals by rivers into the sea, the crustal movements and frequent igneous activities in Eastern China, rich metal and non-metal deposits have been formed in China. These mineral-containing rocks are weathered into clastics which then enter the sea and settle to form a placer belt with various minerals and resources in the coastal zone under the effect of currents and tides.

China has more than 65 kinds of minerals and 191 placer deposits in nearshore coasts, most of which are non-metal placer deposits. 13 minerals have been proven to have industrial reserves, including ilmenite, rutile, zircon, xenotime, monazite, magnetite, alluvial tin and chromite, mainly distributed in coastal areas of Liaodong Peninsula, Shandong Peninsula, Fujian Province, Guangdong Province, Hainan Province and Guangxi Province. Although Hebei, Jiangsu and Zhejiang provinces have a small number of ore spots and anomaly areas, they cannot form industrial deposits due to low grade.

In terms of industrial minerals, monazite, xenotime, ilmenite, rutile, cassiterite, tantalum and tantalum-niobium ores are mainly distributed in Guangdong, Guangxi and Hainan provinces; zircon, quartz sand and gravel are distributed in the coastal provinces; alluvial gold is distributed in Liaoning, Shandong and Taiwan; and diamond is usually found in Liaoning Province.

Minerals in the International Seabed Area

The international seabed area (the "area") refers to the seabed and ocean floor and the subsoil thereof, beyond the limits of national jurisdiction. The mineral resources in the "area" generally mean all the solid, liquid or gaseous mineral resources in the seabed and ocean floor and the subsoil thereof, mainly including polymetallic nodules, cobalt-rich crusts, and polymetallic sulfides. The seabed and ocean floor and the subsoil thereof, beyond the limits of national jurisdiction, as well as the resources therein are the common heritage of mankind. The exploration and exploitation should be carried out for the benefit of all mankind. Based on the UNCLOS, the International Seabed Authority manages, on behalf of the mankind, the resources in the "area". The International Seabed Authority has developed the Regulations on Prospecting and Exploration for Polymetallic Nodules in the "Area", the Regulations on Prospecting and Exploration for Polymetallic Sulphides in the "Area", and the Regulations on Prospecting and Exploration for Cobalt-rich Crusts in the "Area" to manage related activities in the "area".

Polymetallic Nodules

Polymetallic nodules, also called manganese nodules, are a form of deep-sea mineral resources. They are rock concretions on the sea bottom formed of concentric layers of iron and manganese hydroxides around a core. Polymetallic nodules are widely distributed in the depth of 4-6 kilometers under the sea, and contain 70 kinds of elements, with a total

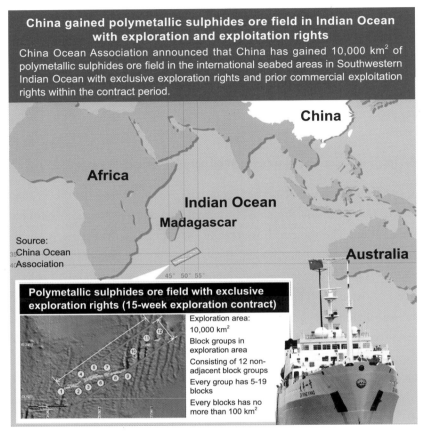

In 2011, the Council of ISA approved application for a polymetallic sulphides ore field put forward by the China Ocean Mineral Resources R & D Association and China gained 10,000 km² of polymetallic sulphides ore field in the Southwestern Indian Ocean.

amount of 3 trillion tons, of which 75 billion tons have the potential for commercial exploitation.

The Pacific Ocean is where polymetallic nodules of the highest economic value are widely distributed. They have a zonal distribution in the Pacific Ocean, mainly distributed in the Northeast Pacific Basin, Central Pacific Basin, Southern Pacific Basin, and Southeast Pacific Basin.

Polymetallic nodules distributed in the Clarion-Clipperton fracture zone of the Northeast Pacific Ocean, the Peru Basin of the Southeast Pacific Ocean, and the center of the North Indian Ocean are of the highest economic value.

In 2001, China Ocean Mineral Resources R&D Association signed a contract with the International Seabed Authority for exploration of polymetallic nodules in the Northeast Pacific Ocean, to obtain the exclusive right to explore the mines of 75,000 square kilometers in the Clarion-Clipperton fracture zone of the central Eastern Pacific Ocean, and to have a priority right to develop polymetallic nodules when entering the commercial exploitation period. There are about 420 million tons of dry polymetallic nodules in mines, containing 111.75 million tons of manganese, 4.06 million tons of copper, 5.14 million tons of nickel, and 980,000 tons of cobalt.

Cobalt-rich Crusts

Oxide deposits of cobalt-rich crusts are distributed throughout the world's oceans. They are widely distributed on the slopes of ocean basins or at the top of flat-top seamounts, usually formed 400-4,000 meters underwater. Thicker crusts and crusts containing richer cobalt are located in the ocean floor 800-2,500 meters deep. The potential crust resources contain up to one billion tons of cobalt metal. Cobalt-rich crusts contain not only cobalt metal, but also other metal and rare earth elements such as titanium, cerium, nickel, platinum, manganese, phosphorus, thallium, tellurium, zirconium, tungsten, bismuth and molybdenum.

The Pacific Ocean, the Indian Ocean and the Atlantic Ocean are where cobalt-rich crusts are concentrated. The crust mine sites with the largest mining potential are located in the Central Pacific Ocean near the equator, especially the Johnston Island, Hawaiian Islands, Marshall Islands, Exclusive Economic Zone around the Federated States of Micronesia, and the international seabed area of the Central Pacific Ocean.

China started late in the survey of cobalt-rich crusts. In 1987, the "Ocean IV" got the cobalt-rich crust sample for the first time. Since 1997, China has conducted a number of strategic explorations for cobalt-rich crusts in the central and western Pacific Ocean. In July 2012, the International Seabed Authority approved the Regulations on Prospecting and Exploration for Cobalt-rich Crusts in the "Area". After that, China submitted to the International Seabed Authority the application for exploring a cobalt-rich crust mining area of 3,000 square kilometers in the international seabed area of the Northwest Pacific Ocean. On July 19th, 2013, the application was approved by the International Seabed Authority Council, which indicated that China had made substantive progress in gaining the right to explore potential strategic resources in the seabed.

Volcanogenic Massive Sulfide Ore Deposits

Volcanogenic massive sulfide ore deposits, also called polymetallic sulfides, are another new kind of heavy metal mineral resources under the sea, following polymetallic nodules and cobalt-rich crusts. Polymetallic sulfides, containing elements such as zinc, lead, gold, and silver, are discharged from the "black chimneys" as seawater enters into stratum space and is heated by lava beneath Earth's crust. The temperature of such hydrothermal fluid is up to 400 ℃. When the hydrothermal fluid is discharged from the stratum space on the seabed and mixed with the surrounding cold water, the metal sulfides in the water will precipitate to the "black chimneys" and the stratum surface under the sea, forming sulfide minerals. Polymetallic sulfide ore deposits are usually found on the seabed with frequent tectonic activities, such as on the mid-ocean ridges, back-arc basins, and intraplate volcanos. There are more than 100 proven hydrothermal mineralization points, of which 25 points have high-temperature black chimney vents, mainly distributed in the Pacific Ocean, the Atlantic Ocean and the Red Sea. According to preliminary estimates, the total resources in the proven hydrothermal mineralization points are up to 400 million tons.

The development of polymetallic sulfides under the sea is only in the exploration stage. Given that the prospecting and exploration for polymetallic sulphides and other seabed mineral resources will have impacts on the marine environment, and that the seabed and ocean floor and subsoil thereof beyond the limits of national jurisdiction and the regional resources are the common heritage of mankind, the International Seabed Authority approved and passed the Regulations on Prospecting and Exploration for Polymetallic Sulphides in the "Area" on the 16th Council held in Kingston, Jamaica from April to May 2010, which specified the detailed provisions on prospecting and exploration of polymetallic sulphides.

Volcanogenic massive sulfide ore deposits have huge potential economic value and good development prospects. China's "Ocean I" has found many polymetallic sulphide ore deposits with commercial exploitation value in the Pacific Ocean and the Indian Ocean. In July 2011, the International Seabed Authority Council approved the application

On May 29th, 2014, China's "Ocean One" scientific research ship returned to the home port of Qingdao after completing its 30th expedition voyage. This was the first voyage of exploration in the Southwestern Indian Ocean since China Ocean Mineral Resources R & D Association and ISA signed agreement in 2012. The picture showed the basalt samples collected from the deep sea.

submitted by the China Ocean Mineral Resources R&D Association for exploration of polymetallic sulphides in the southwest Indian Ocean. The China Ocean Mineral Resources R&D Association therefore gained the exclusive right to explore and the priority to develop the undersea mines of about 10,000 square kilometers.

> Volcanogenic massive sulfide ore deposits in the southwest Indian Ocean are mainly distributed on the ridge of the southwest Indian Ocean, and limited within a rectangular of 990 kilometers long and 290 kilometers wide. The term of the exploration contract is 15 years. Within 10 years after the signing of the contract, China will complete abandonment of 75% of the exploration area, and keep 2,500 square kilometers of area as the mine with exploitation priority.

Valuable Offshore Space

Oceans provide humans with space not only for shipping, fishing, and aquaculture, but also for construction of emerging marine projects required by humans, such as maritime cities, maritime plants, maritime entertainment, undersea tunnels and subsea storehouses. Oceans provide broad space for the future development of humans. The marine space resources refer to geographic areas related to marine development and utilization near, on and at the bottom of the sea. The typical ways to utilize marine space include land reclamation from the sea, aquaculture, construction of port terminals, establishment of waterways, and construction of marine recreational facilities and storage bases.

International Container Terminal in Ningbo Port with numinous cranes.

Over 150 Seaports

China now has more than 150 coastal ports (including the ports in Nanjing and lower reach of the Yangtze River), mainly distributed in the five port clusters of the Bohai Sea, the Yangtze River Delta, the southeast coast, the Pearl River Delta and the southwest coast. (1) The port cluster of the Bohai Rim is composed of coastal ports in Liaoning Province, Tianjin City, Hebei Province and Shandong Province, serving the social and economic development in coastal and inland areas in Northern China. (2) The port cluster of the Yangtze River Delta relies on Shanghai International Shipping Center and is based in Shanghai, Ningbo City and Lianyungang City, serving the economic and social development in areas around the Yangtze River Delta and along the Yangtze River. (3) The port cluster of the southeast coast is based in Xiamen City and Fuzhou City, including ports in Quanzhou City, Putian City and Zhangzhou City, serving the economic and social development of the inland provinces of Fujian and Jiangxi, and meeting the needs of "Three Direct Links" (links of trade, travel and post) with Taiwan. (4) The port cluster of the Pearl River Delta is composed of ports in the eastern Guangdong Province and the Pearl River Delta, based in Guangzhou City, Shenzhen City, Zhuhai City and Shantou City, serving part areas in South China and southwest China, and strengthening the exchanges between inland areas and Hong Kong and Macao. (5) The port cluster of the southwest coast is composed of ports in western Guangdong Province, Guangxi Province, and Hainan Province, based in Zhanjiang City, Fangcheng City, and Haikou City. Now ports in Beihai, Qinzhou, Yangpu, Basuo and Sanya are developing to serve the development in the western region, and providing transport security for Hainan to expand exchange of goods with other areas outside Hainan Island.

Coastal Ports with Container Throughput of over One Million TEUs (Unit: million TEUs)			
Port	Container Throughput	Port	Container Throughput
Port of Shanghai	3252.94	Yingkou Port	485.10
Shenzhen Port	2294.13	Port of Yantai	185.05
Ningbo - Zhoushan Port	1617.48	Port of Fuzhou	182.50
Guangzhou Port	1454.74	Rizhao Port	174.92
Qingdao Port	1450.27	Port of Quanzhou	169.70
Tianjin Port	1230.31	Dandong Port	125.05
Dalian Port	806.43	Shantou Port	125.02
Xiamen Port	720.17	Humen Port	110.36
Lianyungang Port	502.01	Port of Haikou	100.01

Coastal Ports with Cargo Throughput of over One Hundred Million Tons (Unit: hundred million tons)

Port	Cargo Throughput	Port	Cargo Throughput
Ningbo - Zhoushan Port	7.44	Shenzhen Port	2.28
Port of Shanghai	6.37	Port of Yantai	2.03
Tianjin Port	4.77	Beibu Gulf Port	1.74
Guangzhou Port	4.35	Lianyungang Port	1.74
Qingdao Port	4.07	Xiamen Port	1.72
Dalian Port	3.74	Zhanjiang Port	1.71
Tangshan Port	3.65	Huanghua Port	1.26
Yingkou Port	3.01	Port of Fuzhou	1.14
Rizhao Port	2.81	Port of Quanzhou	1.04
Qinhuangdao Port	2.71		

Chinese coastal ports have 1,517 berths over 10,000 tons level, with an annual passenger throughput of 115 million people, cargo throughput of 6.88 billion tons, and container throughput of 158 million TEUs (twenty foot equivalent unit). There are 19 coastal ports with cargo throughput of more than one hundred million tons, and 18 coastal ports with container throughput of over one million TEUs.

More than 160 Large Bays

Bordering on the Bohai Sea, the Yellow Sea, the East China Sea, the

South China Sea and the waters in east of Taiwan, the mainland China has over 160 bays, each covering more than 10 square kilometers, and over 10 large and medium-sized estuaries. The major bays include the Bohai Bay, Liaodong Bay and Laizhou Bay in the Bohai Sea, Hangzhou Bay in the East China Sea, and Beibu Gulf in the South China Sea.

Bohai Bay is a shallow bay in the west of the Bohai Sea, surrounded by land on three sides, adjacent to Hebei, Tianjin and Shandong, and connecting the Bohai Sea on the line between the Luan River Estuary and the Yellow River Estuary. It covers an area of 15,900 square kilometers, accounting for one-fifth of the Bohai Sea. The undersea terrain gets deeper from the land to the central bay, with an average depth of 12.5 meters.

Liaodong Bay is a bay of the highest latitude among all bays in China, located in the northern part of the Bohai Sea and to the north of the line between Changxing Island and Qinhuangdao, as graben-type depression.

Major Bays in China		
Name of Bay	**Province**	**Sea Area**
Liaodong Bay	Liaoning	Bohai Sea
Bohai Bay	Tianjin, Hebei	Bohai Sea
Laizhou Bay	Shandong	Bohai Sea
Jiaozhou Bay	Shandong	Yellow Sea
Haizhou Bay	Shandong, Jiangsu	Yellow Sea
Hangzhou Bay	Zhejiang	East China Sea
Beibu Gulf	Guangxi	South China Sea
Xinghua Bay	Fujian	East China Sea
Sansha Bay	Fujian	East China Sea
Daya Bay	Guangdong	South China Sea

The terrain of the bay tilts from the bay head and the eastern and western sides to the central bay, with the eastern side being deeper than the western side, and the maximum depth of 32 meters.

Laizhou Bay, located in the south of the Bohai Sea, is a bay under the control of the Tancheng-Lujiang fault zone, and is formed by the depressions of the fault zone. It originates from Longkou City in the east and terminates in the old Yellow River Estuary. Due to the accumulation of river sediments, the water depth of the bay is less than 10 meters.

Hangzhou Bay, a typical funnel-shaped bay, starts west from the Ganpu-Xisan gate section and ends east on the Yangzi-Zhenhai line. The width of bay mouth is 100 kilometers, decreasing from outer mouth to inner mouth and to only 20 kilometers in Ganpu. On the north shore of the bay is the southern edge of the Yangtze River Delta, with deep grooves along the shore; on the south shore of the bay lies the Ningshao Plain, with wide beaches along the shore.

Beibu Gulf, located in the north of the South China Sea, is defined to the east by Leizhou Peninsula and Qiongzhou Strait, to the southeast by Hainan Island, in the north by Guangxi, and in the west by Vietnam. It covers an area of 44,238 square kilometers, with the general water depth of 20-50 meters. The maximum water depth is not more than 90 meters.

More than 7,300 Islands with Area of over 500 Square Meters

China owns numerous islands in the sea, including more than 7,300 islands with each covering an area of over 500 square meters. All the islands in China cover a total land area of nearly 80,000 square kilometers and a total coastline of over 14,000 kilometers long. There are abundant biological resources on the islands. The inshore islands have about 2,000 species of plants, of which over 1,000 species are of medicinal value. But due to the small amount of resources, only some 200 species exist in a certain number

The Putuo Mountain, together with Mount Wutai in Shanxi, Mount Jiuhua in Sichuan and Mount Emei in Anhui, is regarded as the four Buddhist Mountains. It is one of the 1,390 islands in Zhoushan Islands, shaping like a dragon lying at sea and known as the "Sacred Buddhist Mountain at Sea".

and scale. On the islands, birds are of the largest number among all animals. They total about 400 species, of which 80% are migratory birds or travelling birds. In the south of Fujian Province, especially near the Hainan Island, there are rich resources of mangroves and coral reefs.

Chinese islands have a small population with concentrated distribution. At present, China has 2 island municipalities, 14 island counties (cities and districts), and 191 island townships (towns). In 2007, the population of all islands in China was about 5.47 million (excluding Hong Kong, Macao, Taiwan and Hainan Island), of which 98.5% live in the above central island cities, counties and townships. The economic output of the islands is small, with single economic structure. The output value of the marine fishery industry generally accounts for a larger proportion of the GDP on the islands. Uninhabited islands are used for various purposes. There are over 1,900 uninhabited islands China has utilized, including 1,020 islands for special purposes, 365 islands for public services, 73 islands for tourism and entertainment, 340 islands for farming, forestry, animal husbandry and fishery, 49 islands for industry, warehousing and transportation, and over 80 islands for renewable energy and urban and rural construction. Islands in the sea and the surrounding waters are rich in biological resources both in quantity and in variety. The wide intertidal zones and nearby waters are places for various species of fish, decapod, shellfish, and algae to spawn, breed, feed and inhabit because there are a large number of economic aquatic biological resources available for eating, curing and aquaculture.

In April 2012, the National Island Protection Plan (2011-2020) was formally announced for implementation, which was China's major initiative in promoting the development of the islands. Meanwhile, China also announced the first batch of 176 uninhabited islands available for development and utilization, involving eight provinces including Liaoning, Shandong, Jiangsu, Zhejiang, Fujian, Guangdong, Guangxi, and Hainan. Among the 176 uninhabited islands, 11 are in Liaoning, 5 in Shandong, 2 in Jiangsu, 31 in Zhejiang, 50 in Fujian, 60 in Guangdong, 11 in Guangxi, and 6 in Hainan. According to the Measures for the Administration of Collection and Application of Use Fees for Uninhabited Islands, the right to use uninhabited islands can be transferred by ways of application approval, bidding, auction and quotation.

More than 100 Beautiful Coastal Beaches

China has over 1,500 seaside tourist attractions and more than 100 coastal beaches with beautiful natural landscapes. The State Council has announced 16 national historical and cultural cities, 25 national key scenic

The beautiful seashore landscape in the Dadonghai Bay in the city of Sanya, Hainan. People are sightseeing and resting along the beach in the shadows of coconut trees.

spots, and more than 210 marine protected areas such as typical marine ecosystems, rare and endangered marine life, natural and historical sites and natural landscapes, including 32 national marine nature reserves, 17 national marine special reserves, and 57 marine protected areas related to the islands.

Chinese marine tourism relies on the development in coastal cities, to form the four coastal tourism belts of the Bohai Sea, Yangtze River Delta, Pearl River Delta and Hainan Island. The coastal tourism belt of the Bohai Sea is based in Dalian City, Qinhuangdao City and Qingdao City, and is a representative of China's northern waterfront. The coastal tourism belt of the Yangtze River Delta is based in Shanghai and supplemented by tourist cities of Lianyungang, Ningbo, Hangzhou, Nanjing, and Suzhou, forming a coastal tourist belt with link between land and sea, and integration of cultural and natural landscapes. The Pearl River Delta, one of China's most economically developed regions, has formed a tourism belt based in Hong

Kong, Macau, Guangzhou City, Shenzhen City, Zhuhai City, Shantou City, Zhanjiang City, and Beihai City. Hainan Island is China's most famous coastal tourism area, owning the world-class tropical coastal tourism resources. Hainan is trying to build itself into an international tourist island, and has significantly accelerated the pace of development of its tourist resources.

CHINA'S MARINE CONSERVATION AND DEVELOPMENT

"VITAMIN" TO CHINA'S 1.3 BILLION POPULATION
——CHINA'S UTILIZATION OF MARINE RESOURCES

The global marine resources initially proven to be available for development

230,000 kinds of marine biological resources

More than **800** kinds of important fishing targets

Consisting of **19,000** kinds of fish

200-300 million tons of total allowable catches

About **135** billion tons of recoverable oil reserves

About **140** trillion cubic meters of recoverable natural gas reserves

98% of the global gas hydrates are stored in the ocean

the carbon content of which is twice the carbon content of global fossil fuels

The renewable energy resources in the ocean are approximately **7** billion kilowatts

which is ten times of the world's current power generation capacity

97.3% of global water resources lie in the ocean

Sea water resources are unlimited resources

It is estimated that the number of mineral and biological resources in the ocean is 1,000 times of that on the land, and about 85% of the species on Earth are living in the ocean. In addition, there are abundant metallic mineral resources and deep-sea biological genetic resources in the international waters with great potential commercial exploitation value.

China is a big country with both land and sea, vast waters under its jurisdiction, and rich coastal and marine resources. Currently, the marine resources available for development and utilization mainly include marine biological resources, mineral resources, sea water resources, ocean space and marine renewable energy. The development of marine resources has formed more than 10 marine industries, which creates the added value in marine industries, accounting for about 10% of China's GDP. Marine resources development activities have also provided more than 33 million jobs.

No. 1 in Mariculture Production in the World

Marine fishing and mariculture is an important form to develop living marine resources and provide marine products for coastal and inland areas, and also an important source of animal protein for humans. In 2012, China's seafood production reached 30.33 million tons, accounting for 51.3% of the total output of China's aquatic products and representing a YoY increase of 4.31%.

In the latter part of the 20th century, China's fishing of offshore fishery resources grew rapidly. In 1995, the harvest reached more than 10 million tons, ranking first in the world. But since 1999 when China implemented offshore fishing "zero growth" strategy, the harvest of offshore fishery resources has maintained about 13 million tons per year. China has strengthened the control of the number and power of fishing boats and nets, improved the fishing license system, actively guided the restructuring of offshore fishing production, and further reduced the use of trawls and stow nets that had greater impact on resources, so that the production structure in the offshore fishing industry becomes increasingly reasonable. The orderly development of coastal pelagic fish resources and the better use of demersal resources have facilitated the gradual optimization of the product structure. Meanwhile, China has accelerated the transfer of coastal fishermen into other industries, to effectively improve the re-employment of fishermen through the implementation of job transfer.

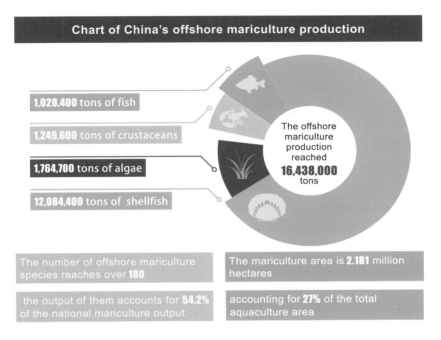

Speaking of tonics, people may think of sea cucumbers in the Northeast China, especially in Dalian City. Dried, fresh, and powdered sea cucumbers and sea cucumbers made into capsules are sold in various markets and even in malls throughout China. But wild sea cucumbers are very few. Most of the sea cucumbers are artificially farmed.

China is a big mariculture country in the world. In 2012, the offshore mariculture production reached 16,438,000 tons, of which 1,028,400 tons were fish, 1,249,600 tons were crustaceans, 12,084,400 tons were shellfish, and 1,764,700 tons were algae. The number of offshore mariculture species reaches over 180 and the output of them accounts for 54.2% of the national mariculture output, and 80% of the global mariculture output. The mariculture area is 2.181 million hectares, accounting for 27% of the total aquaculture area, where aquaculture resources in Liaoning Province, Zhejiang Province, Fujian Province, Shandong Province, Guangdong Province, Jiangsu Province and Guangxi Province account for 70% of

China's total. With the implementation of advantageous aquaculture area development plan, there are now advantageous processing areas for Penaeus vannamei Boone, eel, squid, tilapia, crayfish and surimi.

China's mariculture space is expanding, from traditional pond culture, beach culture, near shore culture to offshore culture. Mariculture facilities and equipment continue to increase. Factory farming and cage culture maintain sustainable development, and the degree of mechanization and automation has improved significantly. The level of industrialization has continuously been improved. The socialization and organization of the mariculture industry has significantly enhanced, forming an industry group integrating seed cultivation, seed breeding, feed production, mechanical support, and standardized breeding, processing and marketing.

Complete Category of Marine Product Processing Industry

China's marine product processing industry develops very fast, facilitating the continuous increase of marine product processing capabilities. By 2011, the total number of marine products processed reached 14.78 million tons, accounting for 82.7% of total number of aquatic products processed. Through continuous improvement of processing technologies, China has formed dozens of industrial categories, including refrigerated marine products, pickled marine products, smoked marine products, powdered fish, seaweed food, and marine drugs, and established a number of listed companies and well-known enterprises in the marine product processing industry.

Most marine product processing enterprises gather in the places where the raw materials are, which helps them realize the optimization of raw material supply, storage, and transportation costs. In areas like Dalian of Liaoning Province, Qingdao, Yantai, Weihai and Rizhao of Shandong Province, Zhoushan, Ningbo, Wenzhou and Taizhou of Zhejiang Province, Fuzhou, Xiamen, Zhangzhou and Ningde of Fujian Province, Zhanjiang, Shantou and Chaozhou of Guangdong Province, Yancheng and Nantong of Jiangsu Province, and Beihai of Guangxi Zhuang Autonomous Region, there are obvious processing advantages. With the support policies issued by central and local governments, the above areas have the advantages of raw material concentration and the marine product processing industry shows a cluster development trend.

Four Marine Biotechnology and Drug Research Centers

Currently, there are about 1,000 species of marine organisms known for their medicinal use in the world, from which hundreds of natural products are isolated and of which more than ten kinds of single drugs and nearly 2,000 kinds of compound Chinese patent medicines are made. In addition, we have also found in marine organisms more than 10,000 kinds of compounds with new structures, and applied for a patent for more than 200 kinds of them.

Since 1985 when China developed its first marine drug Polyalginate Sulfate Sodium (PSS), the development and research of marine organisms in China has achieved fruitful results. China has developed a number of marine medicines to cure major diseases of serious harm to human health, and has effectively promoted faster development of China's marine pharmaceutical industry.

In 2011, China's marine biomedical industry achieved the added value of RMB 17.2 billion. The marine biomedical research has been gradually standardized, with formation of four marine biotechnology and marine drug research centers based in Shanghai, Qingdao City, Xiamen City, and Guangzhou City respectively. In coastal areas, there are tens of institutions engaged in research of natural marine medicines. Shandong, Guangdong, Jiangsu, Fujian and other maritime provinces have also stepped up their investment in the marine bio-pharmaceutical industry, and taken the industry as a major economic growth point.

"Going Global" and "Inviting In" for the Exploitation of Oil and Gas Resources

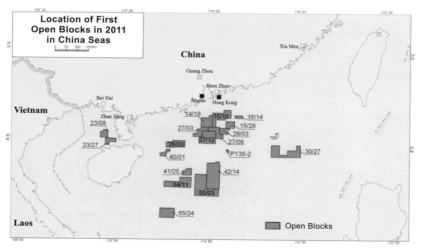

Location of First Open Blocks in 2011 in China Seas

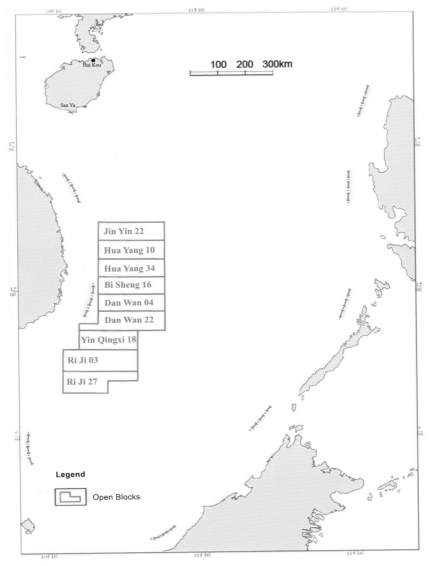

Location of First Open Blocks in 2012 in China Seas

In recent years, China has accelerated the pace of exploration and development of oil and natural gas resources. The oil and natural gas

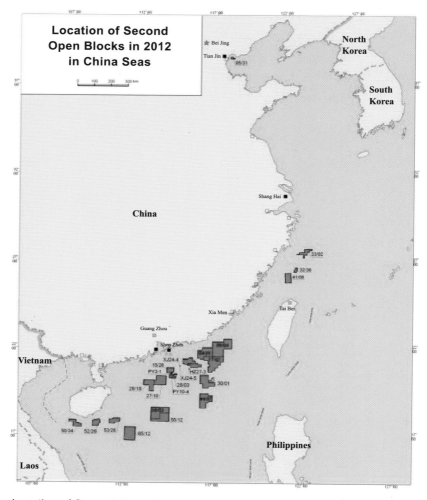

Location of Second Open Blocks in 2012 in China Seas

industry develops fast. In 2012, a total number of 1,376 wells were drilled in domestic and international seas, with 1,055 wells completed. The same year witnessed a global production of 51.86 million tons of crude oil, 16.4 billion cubic meters of natural gas, and 470 million cubic meters of coalbed methane (CBM). The year of 2012 also saw a domestic production of 38.57 million tons of crude oil and 11.3 billion cubic meters of natural gas, accounting for nearly 20% of China's annual production of oil and natural gas. Throughout 2012, China realized an annual production of 50 million tons of oil and natural gas within China. In the whole year, China processed 30.08 million tons of crude oil and produced 7.79 million tons of refined oil.

China's exploration of marine oil and natural gas is mainly concentrated in the Bohai Sea, the East China Sea, the South China Sea and other inland seas and offshore shallow waters. In recent years, while implementing the "Going Global" strategy, China has also adopted the "Inviting In" strategy to increase international cooperation and has announced a number of bidding blocks for deep-sea oil and gas resource development. In May 2011, China introduced 19 oil and gas bidding blocks in the northern South China Sea and the Beibu Gulf, with a total area of 52,006 square kilometers. In June and August 2012, China launched 9 and 26 oil and gas bidding blocks in the South China Sea respectively, with a total area of 233,878 square kilometers and a total of seven oil contracts signed, achieving new results in foreign cooperation.

Integration and Industrialization of Seawater Utilization

There are 80 kinds of natural elements in seawater, including oxygen, hydrogen, chlorine, sodium, magnesium, sulfur, calcium and potassium, all of which are of a high content. The production of salt, magnesium and bromine from seawater, comprehensive utilization of brine, and extraction of potassium, iodine, and uranium from seawater can largely compensate for the shortage of land resources.

The seawater utilization in China, especially in coastal cities

Technical personnel were discussed installation of equipment on the construction site of the Yancheng new energy desalination demonstration project.

suffering from severe water shortage, is heading towards recycling and industrialization. In modern technical and economic conditions, industrial water, domestic water, and water for irrigation of salt-tolerant plants can be directly obtained from the sea. Seawater can even solve part of the drinking water problem after it is desalinized. To strengthen the development and utilization of seawater resources is an important measure to solve freshwater crisis and water shortage in coastal areas.

Sea Salt Resources

China ranks top in the world in terms of its sea salt output. In 2011, the output value of China's sea salt industry reached RMB 7.68 billion, accounting for 0.4% of that of China's marine industry. China's sea salt output reached up to 33.22 million tons, with the highest yield falling in Shandong Province, which was 22.73 million tons, accounting for 70% of China's national output.

> Changlu Salt Field is mainly located on the coast of the Bohai Sea between Shandong Province and Tianjin Municipality. The largest salt pan is Tanggu saltpan, with an annual salt output of 1.19 million tons. Changlu Salt Field originates south from Huanghua and extends north to southern Shanhai Pass, including salt pans of Tanggu, Hangu, Dagu, Nanbao and Daqinghe River, with a total length of 370 kilometers, a total area of over 2.3 millionmu, and an annual output of more than 3 million tons of sea salt.
>
> Budai Salt Field is the largest salt field in Taiwan. It is located in the southwestern coast of Taiwan Island, with an annual output of over 600,000 tons of table salt, and is thus known as the "Southeast Salt Warehouse."
>
> Yinggehai Salt Field, located in the southwestern shore of Ledong, is the largest sea salt field on the Hainan Island and in southern China. Built in 1958, the salt field covers a total area of 3,793 hectares, with an annual production capacity of 250,000 tons of salt.

Seawater Desalination

Desalination refers to the process of removing salt from sea water. It is a technology to increase water resources. Desalination can increase the amount of fresh water which is of good quality and at reasonable prices regardless of time and climate, and can ensure stable supply of drinking water for coastal residents and of boiler feed water for industries.

From the beginning of the 1990s, with the increasingly serious water shortage situation in China, desalination has entered a period of great development and gradually moved toward large-scale applications. After 40 years of development in seawater utilization in China, the main process of desalination technology has become relatively mature and presented the unitization and modular tendency.

China has built and is building desalination projects with a total capability of desalinating about 600,000 tons of water per day. The

The staff of the Desalination Workshop of the Datang Huangdao Power Generation Co., Ltd, Shandong Porvince, was monitoring desalinated water quality.

industrialization demonstration projects already built include reverse osmosis desalination projects with a capacity of desalinating 5,000 tons of water per day, low-temperature multi-effect distillation desalination projects with a capacity of desalinating 3,000 tons of water per day, and low-temperature multi-effect desalination projects with a capacity of desalinating 12,500 tons of water per day. Through the introduction, digestion, absorption and innovation of foreign advanced technologies, China has managed to design and manufacture complete sets of equipment for low-temperature multi-effect desalination of 3,000 tons and 4,000 tons of water per day.

Direct Seawater Utilization

Direct seawater utilization refers to using sea water instead of fresh water as the industrial, agricultural, commercial and domestic water to alleviate the shortage of freshwater resources in coastal areas. Industrial water mainly means cooling water. Agricultural water refers to water that can be directly used for irrigation in agricultural production and mariculture industries in coastal areas. Urban domestic water includes water for toilet rush and firefighting. Direct seawater utilization is a technical means to directly use sea water to replace and save fresh water, promoting the optimization of the water resources structure.

The amount of seawater directly used in China is over 60 billion cubic meters. Mainly used for industrial cooling and urban life, the seawater has become an important source of water resources in coastal areas. Although seawater utilization accounts for only a small proportion in the total marine economy, its real significance cannot be measured only by industrial output. Seawater utilization can not only increase water resources, but also play a long-term and strategic role in supporting other industries and guaranteeing water security.

China's coastal cities have over 60 years of history of direct use of seawater as industrial cooling water. Power plants in coastal areas and petrochemical enterprises in Dalian City, Qingdao City, Tianjin City and Shanghai all use seawater as cooling water in their industrial processes of oil refining, chemical fiber production, soda manufacturing, acid making, ammonia synthesis, oleo chemicals and dye, and have achieved tremendous social and economic benefits.

Urban domestic water refers to using seawater as urban domestic water to save about 35% of the domestic water. It has important social and economic benefits, and broad application prospects. Hong Kong SAR of China has almost 50 years of history of using seawater as residents' flushing water and has formed a complete processing system and management system. At present, 76% of the Hong Kong population use sea water for flushing, with an annual water consumption of 200 million cubic meters, which accounts for about 18% of the total daily water consumption in Hong Kong.

Seawater irrigation refers to using seawater to irrigate crops. Large-scale plantation of seawater resistant crops on beaches can create land with silt and slow down the erosion of coastal land by seawater. Meanwhile, it can, to some extent, alleviate the pollution caused by industries and aquaculture on coastal beaches and in offshore areas, mitigate the greenhouse effect, and improve the ecological environment. At present, the seawater resistant crop of Salicornia bigelovii Torr is gradually introduced and planted in China's coastal areas.

Utilization of Marine Renewable Energy

Marine renewable energy resources mainly include marine tidal energy, marine wave energy, ocean current energy, ocean salinity energy, ocean thermal energy, and marine wind energy. The utilization of marine renewable energy is featured by light pollution, small space occupation, and less pressure on consumption of fossil fuels. Besides, it has comprehensive utilization value. For example, seawater can not only generate power, but also be used for aquaculture and convenient transportation. The places for wind power generation can also be used to develop sea ranches and coastal tourism.

Tidal Energy

China's tidal energy resources reserves in coastal areas are up to 1.1×10^8 kilowatts, with an installed capacity available for use being about 35 million kilowatts. China's tidal energy resources are mainly concentrated along the coast of the East China Sea, most of which are distributed along the coasts of Fujian and Zhejiang provinces. In terms of the energy density and geological conditions of bays, the coasts of Fujian, Zhejiang provinces are the best places to develop China's tidal energy resources, followed by

Jiangxia Testing Tidal Power Station, Asia's largest tidal power station

the east side of the south bank of the Liaodong Peninsula, the north side of the south bank of the Shandong Peninsula and the eastern Guangxi Province. With large tidal ranges, those places are bedrock embayment coasts, favorable for constructing tidal power stations.

The tidal energy development and utilization technology is the most mature technology among all technologies for marine renewable energy development and utilization, where China is also experienced in long-term operation, management and maintenance of the power stations. The total installed capacity of tidal power stations in China is 6,000 kilowatts, ranking third in the world, while that of Jiangsha experimental tidal power station in Wenling City that has operated for over 30 years ranks fourth in the world. At present, China has still more than a dozen of tidal power stations under construction, of which the tidal power station in Jiantiao Bay in Zhejiang Province is planned to have an installed capacity of 20,000 kilowatts.

Wave Energy

Wave energy is the most unstable energy among all marine energies. It is generated when wind transmits its energy to the sea and is essentially formed by absorption of wind energy.

The geographical distribution of China's wave energy is very uneven, with the densest distribution in Taiwan, followed by Zhejiang Province, Guangdong Province, Fujian Province and Shandong Province, with the lowest distribution in Guangxi coast. The power density of wave energy in coastal areas of outlying islands is higher than that in coastal areas of offshore islands. The wave power in coastal areas of offshore islands is stronger than that in mainland coast. The power density of wave energy in coastal areas of the Bohai Strait, northern and southern Taiwan and the Xisha Islands is relatively high.

Tidal Current Energy

The geographical distribution of China's tidal current energy is very uneven, with the densest distribution in Zhejiang coast, accounting for over 40% of China's total amount. In terms of various sea areas, the East China Sea coast has the highest distribution of tidal current energy, accounting for 80% of the total amount of China, followed by the Yellow Sea coast, while the South China Sea coast has the lowest distribution. Hangzhou Bay and Zhoushan Islands are where the power density of tidal current energy is the highest in China. In addition, the power density of tidal current energy is

The Tidal Current Power Station in Haizhou Bay of Lianyungang

also very high in waters surrounding Laotieshan Mountain in the northern Bohai Strait, Sandu Gulf of Fujian Province, and Fisherman Island of Taiwan's Penghu Islands.

Offshore Wind Energy

Offshore wind energy is a clean and renewable energy, mainly used to generate power and provide electricity for the life and production on islands and in offshore oil and gas fields.

China is a country affected by monsoon climate which is characterized by large affected scope and powerful strength, so that China is rich in offshore wind resources and has good market potentials and huge resources for wind power development and utilization. In the offshore area where the water depth is 10 meters, the wind energy resources are about 100 million kilowatts. In the offshore area where the water depth is 20 meters, the wind energy resources are about 300 million kilowatts. In the offshore area where the water depth is 30 meters, the wind energy resources are about 490 million kilowatts. 10 meters over the offshore area where the water depth is 50 meters, the wind energy resources are about 940 million kilowatts.

The Shanghai Donghai Bridge 100 MW offshore wind power demonstration project is Asia's largest offshore wind farm.

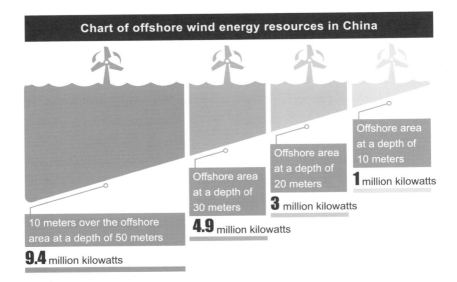

The wind energy resources are mainly distributed in southeast coastal areas. Fujian has the highest wind energy reserves of 211 million kilowatts while Tianjin has the lowest wind energy reserves of only 1.11 million kilowatts.

In 2007, the offshore wind power station in the Suizhong 36-1 oilfield in Liaodong Bay of the Bohai Sea was connected to the electricity supply network and started to supply electricity, marking that the development of China's offshore wind energy entered a substantive start-up phase. This power station was designed, constructed and installed by China independently, with the maximum power output of 1,500 kilowatts at full load. In 2012, all the generators of the 150,000 kilowatts offshore wind farm demonstration project in Rudong County of Jiangsu Province were connected to the electricity supply network and started to supply electricity, becoming Asia's largest offshore wind power station.

Thermal Energy

The average water temperatures of the East China Sea, the Yellow Sea, and the South China Sea are very high. Especially in summers, the average water temperature of the South China Sea is up to 36 ℃. Since most part of the sea is more than 1,000 meters deep, the water temperature at the depth of 500-1,000 meters under the sea is only 5°C, so that it has favorable conditions and broad prospects to use seawater temperature differences to generate power. Due to low latitude, water depth and broad sea area, the South China Sea has abundant thermal energy resources, accounting for over 90% of the total thermal energy in China. The East China Sea and the waters in the east of Taiwan are also rich in thermal energy resources.

Ocean Salinity Energy

The reserves of ocean salinity energy depend on the amount of fresh water into the sea, the salinity of the seawater, and the uneven distribution of the amount of fresh water into the sea. In China, the ocean salinity energy is mainly distributed along the coast of the Yangtze River Delta and around the estuaries of great rivers in the south of the Yangtze River Delta. The installed capacity available for development along the coast of the Yangtze River Delta accounts for over 60% of China's total capacity; while that in the Pearl River Delta accounts for about 20%. The seasonal changes of power of ocean salinity energy are dramatic and its inter-annual variability is also significant. The characteristics in changes of the amount of fresh water into the sea determine the seasonal changes and significant changes of power of ocean salinity energy.

CHINA'S MARINE CONSERVATION AND DEVELOPMENT

POSITIVE ENERGY IN THE FACE OF NATURAL DISASTERS
—— PREVENTION AND REDUCTION OF MARINE DISASTERS IN CHINA

Marine Disasters

In August 2012, "Typhoon Haikui" attacked China, which was the third typhoon that had landed China within 10 days. Under the influence of "Haikui", severe waterlogging appeared in some parts of Zhejiang Province: in Sunhou Village, Xiwu Street, Fenghua, Ningbo City, the water level on the ground reached 1.6 m; around Lianhua Village, Xiangshan, the highest level of water on the road had been over the waist of an average adult male. Landslide and debris flow occurred frequently. In addition, water level on the ground in some of the residential quarters in Shanghai was at the same height of one's waist at worst. Dozens of trains in Nanjing City and all the high-speed trains between Shanghai and Nanjing City and between Nanjing City and Hangzhou City were cancelled.

"Haikui" with accompanying storms affected to varying degrees the four provinces (municipality) of Zhejiang, Shanghai, Jiangsu and Anhui at different levels, leading to 2 deaths and urgent evacuation of 2,086,000 people. 4,800,000 people were stricken by the storm, over 3,700 houses were damaged and 170,900 hectares of crops were hit by the storm, with 17,800 hectares of crops having no harvest, causing direct economic losses in excess of RMB 10 billion.

In 2009, typhoon "Morakot" landed on the coastal area of Hualian, Taiwan. Extraordinary rainstorm brought by "Morakot" broke the historical record of precipitation, leading to severe disasters in Tainan, Gaoxiong, Pingtung and Taitung. The disaster was the most serious one for Taiwan for the last five decades with 673 deaths, 26 missing, NT$ 19.5 billion of agricultural economic loss. The six provinces (municipality)of Zhejiang, Fujian, Jiangxi, Anhui, Jiangsu and Shanghai, were hit to different degrees, where 11,574,500 people were affected, with 12 deaths and 2 missing and direct economic losses up to RMB 12.823 billion.

Located in the west of the Pacific Ocean, China has seen the most frequent occurrence of typhoons and cyclonic activities, and also belongs to one of countries that have suffered from the most frequent and serious marine natural disasters in the world. Especially in recent years, storm surge, coastal erosion, seawater intrusion and other ocean disasters have occurred more frequently as a result of global climate changes. Thus, prevention and reduction of marine disasters have become an important task to ensure sustainable and healthy development of Chinese economy and society.

At present, a stereoscopic marine observation network has taken its initial shape, containing coast earth stations, buoyages, submersible buoys, ships, aeroplains, satellites, radars, etc. Thus, China's marine disaster forecast and mitigation system has been basically formed, and plays an important role in preventing and controlling marine disasters every year.

Storm Surge and Disastrous Wave

Marine natural disasters mean disasters which happen on the sea or coast due to abnormal or violent changes of the marine natural environment, and mainly include storm surge, ocean wave disaster, sea ice disaster, tsunami, etc.

The causes of marine disasters are mainly as follows: intense atmosphere disturbance such as tropical cyclones and extratropical cyclones; disturbance or sudden change of the state of the seawater itself; sea earthquakes, volcanic eruptions and consequent submarine landslides and fractures. Marine disasters not only pose a threat to the security of the sea and coastal areas, but also jeopardize coastal urban and rural economy and people's lives and properties sometimes. Storm surges may give rise to coastal erosion and soil salinity in affected areas; sea earthquakes cause secondary disasters and earthquake-induced disasters such as tsunamis. Marine disasters are anomalous events in geographical environment evolution, and have become one of the most important natural factors that hamper economic and social development in coastal areas.

Storm surges and disastrous waves may occur in any season within a year and in any coastal area from Hainan Island in the south to Liaodong Peninsula in the north. Tropical storms are concentrated in four months, namely, July, August, September and October, especially August and September. Stronger storm surges in the temperate zone mainly happen in late autumn, winter and early spring, i.e. November to April of the following year. Due to global climate changes, marine storm surge disasters gradually move toward the north, bringing Jiangsu Province, Shandong Province and Liaoning Province under increasing threat of storm surges, and hindering local economic development.

With the development of coastal economy and society and addition of infrastructure, the scale of the storm surge disaster is on the rise, and direct and indirect economic losses also grow with years. Direct economic losses brought by storm surge disasters in China increased from the annual average of RMB 0.1 billion in 1950s, to near RMB 2 billion in the late 1980s, and finally to around RMB 7.6 billion in the early 1990s. Since 2001, China has suffered from 256 storm surges, with a disaster-stricken population of 123.81 million, 849 deaths or missing and direct economic losses of RMB 143.3 billion. The direct economic losses in 2005, 2006, 2008 and 2012 were RMB 33 billion, RMB 21.7 billion, RMB 19.2 billion and RMB 12.6 billion respectively. The most disaster-stricken areas are Hainan Province, Guangdong Province, Fujian Province and Zhejiang Province.

In 2012, 24 storm surges occurred in the coastal areas of China, including 13 typhoon-caused storm surges (9 of which caused damages with direct economic losses of RMB 12.629 billion and 9 deaths or missing) and

> Storm surge is aperiodic, irregular increase or decrease of the sea level as a result of strong wind and sudden pressure changes brought by stopover of tropical storm, extratropical cyclone and hail from sea. Storm surge calamity is defined as the disaster to coastal areas owing to the flood brought by storm surge. All coastal areas of China from south to north are subject to attack of storm surge, and storm surge disasters are often embodied as water increase disasters like land submergence, coastal erosion, channel siltation and dam devastation, while water decrease disasters are reflected in channel blockage, difficult access to water for power plants and inconvenience of loading and unloading at ports.

11 extratropical storms with no harm. Typhoon-caused storm surges were concentrated in the period of time from July 23rd to August 28th, with a total of 7 storm surges in coastal areas while extratropical storms in 2012 were at a record low in the recent 5 years, only once over the local warning water level.

As of 2012, China started to carry out the new "Specification for Warning Water Level Determination". Following the new specification, governments in coastal areas check their warning water levels to strengthen the marine disaster prevention and mitigation system. In order to detail the requirements on all aspects of the check, China also issued the "Administrative Measures for Warning Water Level Determination", according to which coastal warning levels shall be updated at least every five years and such update interval shall be shortened appropriately in light of local conditions of different areas.

In accordance with the "Emergency Response to Storm Surge, Tsunami and Sea Ice Disasters", Chinese government issued a red alert for the storm surge upon its occurrence. Governments of Zhejiang and Fujian provinces started emergency responses by spreading disaster information through multiple channels and in different ways such as news, radio and TV and effectively arranging fishing boats and merchant ships back to the port

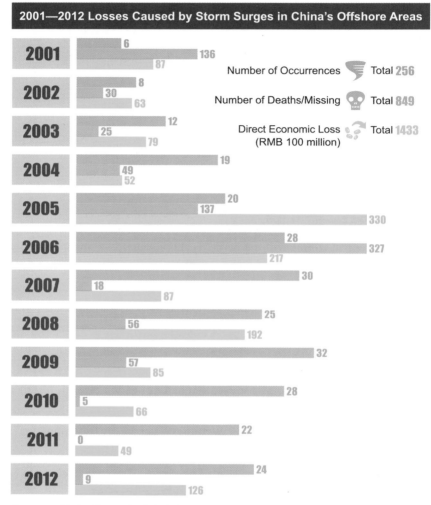

Source: State Oceanic Administration, China Oceanic Disasters Bulletin (2001-2012)

for shelter, evacuating over one million people in the affected areas in time.

After the disaster, Chinese government actively carried out disaster relief and post-disaster production recovery quickly. On the one hand, it conducted marine search and rescue and performed comprehensive search

The village was surrounded by flood after the typhoon.

and rescue in especially seriously stricken waters. On the other, it also exerted great efforts on settlement of the displaced people by effectively dispatching emergency supplies to the affected areas such as tents, foodstuff and clothes and earnestly arranging lives of the stricken people after the disaster. Various local governments and departments organized a large number of laborers, material resources and mechanical equipment to go to all lengths to repair seriously damaged infrastructure.

Disastrous wave in China gradually decreases from the south to the north in terms of the frequency. Its frequency in the South China Sea, the

> Disastrous wave is defined as a disaster where sea wave causes economic losses and causalities including damage and sinking of ships, channel siltation, destruction marine oil production facilities of and coastal projects and impairment of the marine aquaculture industry.

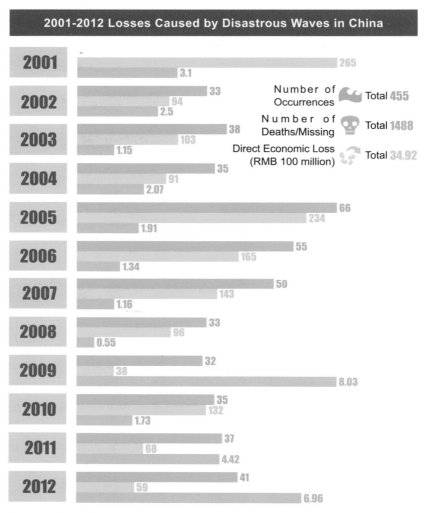

Source: State Oceanic Administration, China Oceanic Disasters Bulletin (2001-2012), 2002-2013.

East China Sea and the Yellow Sea is a bit higher, while that in the Bohai Sea is a bit lower. During the period of 2002-2012, there were a total of 455 disastrous wave disasters and 41 at an annual average. The period from 2005 to 2007 saw the highest occurrence of disastrous wave with an average annual occurrence of 57, and in 2005, the occurrence frequency was as high as 66. The period since 2008 has seen a lower occurrence frequency of disastrous wave. From 2001 to 2012, disastrous wave in China led to 1,488 deaths or missing and direct economic losses of RMB 3.492 billion. Among these years, 2005 saw the highest figures of deaths/the missing with a total number of 234, while recorded the greatest direct economic losses up to RMB 803 million.

In 2012, Chinese offshore areas suffered from 41 disastrous wave processes, including 18 typhoon waves and 23 cold air waves and cyclone waves, which caused 59 deaths (including the missing) and direct economic losses up to 696 RMB million. The direct economic losses were higher than before, and mainly from Liaoning and Shandong provinces, which suffered losses of RMB 448 million and RMB 149 million respectively, accounting for 86% of the total direct economic losses.

> All ice on the sea is named sea ice, which Includes, in addition to ice frozen directly from sea water, ice from terrestrial rivers, lakes and glaciers. Sea ice often results in channel blockage, ship damage and sinking and building damage, which are referred to sea ice disasters collectively. Sea ice disasters can push over marine oil platforms, damage facilities of marine projects and channels, crack up ships, freeze ports and destroy mariculture facilities and sites.

Sea Ice

Sea ice is a main natural hazard in the Bohai Sea, the Yellow Sea and coastal areas. In general, in the "lighter ice year" or "light ice year", sea ice will not produce noticeable effects on sea activities or only exert somewhat effects on water-frozen ports. Despite this, it may bring about sea ice disasters in some areas. Under ice conditions worse than perennial ice years, especially in the much heavier ice years, sea ice will give rise to disasters.

Over the past fifty years, on account of global climate change and other factors, ice period duration and ice class have shown a declined tendency, with no heavier ice years in the Yellow Sea and the Bohai

Major Sea Ice Disasters in China		
Year	Ice Conditions	Impacts or Losses
2011/2012	Perennial ice year	Affected 72,000 people, direct economic losses of RMB 155 million
2010/2011	Perennial ice year	Direct economic losses of RMB 881 million
2009/2010	Heavier ice year	Affected 61,000 people, direct economic losses of RMB 6.32 billion
2008/2009	Lighter ice year[1]	-
2007/2008	Lighter ice year	Offshore oil and gas operations in Liaodong Bay affected
2006/2007	Light ice year	Lightest in the history
2005/2006	Perennial ice year	Severe ice conditions in Laizhou Bay
2004/2005	Perennial ice year	Liaodong Bay was frozen
2003/2004	Lighter ice year	Liaodong Bay, northern Yellow Sea, and Yalu River estuary hard hit
2002/2003	Lighter ice year	Difficult sailing in Liaodong Bay
2001/2002	Light ice year	-
2000/2001	Heavier ice year	Closing of port in northern Liaodong Bay

1. Refers to the average iceconditions during 1978-2008.

In January 2013, the ice conditions in Yellow Sea were more serious than those in previous years, fishing operations and marine aquaculture affected.

Sea. However, the possibility of heavier ice years under abnormal climatic conditions cannot be ruled out. In the periods of 2000-2001 and 2009-2010, 2 heavy ice years happened in Chinese offshore areas, which had been the only two years of this kind in the last 20 years. From January to February in 2010, a sea ice disaster hit the Bohai Bay and northern Yellow Sea, which had been rare in the past 30 years, affecting 61,000 people and causing direct economic losses of RMB 6.32 billion.

In the winters of 2009 and 2010, the Bohai Sea and the northern part of the Yellow Sea went through a heavy ice year, and suffered in the middle and late January of 2011 the most serious ice condition in the past 30 years. From 2009 to 2010, the ice disaster occurred earlier than before, with emergence of a large area of new ice in the Liaodong Bay in late November, half a month ahead than usual. The ice conditions were getting worse at a high speed, as in a short time, the floating ice in the Liaodong Bay extended its scale from 38 sea miles to 71 sea miles, and of the floating ice in the Laizhou Bay grew from 16 sea miles to 46 sea miles, the largest sea ice scale in the last 40 years.

The ice disaster in the winters of 2009 and 2010 in the Bohai Sea and the northern part of the Yellow Sea had serious impact on society and economy of coastal areas and caused huge losses. It affected 61,000 people in the three provinces Liaoning, Hebei and Shandong and the city of Tianjin, destroyed 7,157 ships, froze 296 ports and docks and damaged 207,870 hectares of aquaculture. The direct economic loss was summed up to RMB 6.318 billion.

Sea Level Rise

The rise of the sea level is a slow onset natural disaster, and a global phenomenon caused by global warming, polar ice melting and upper ocean heat expansion. The average sea level from 1975 to 1986 is defined by the international community as the annual mean sea level. The margin of sea level rise is a numerical value by which the sea level rises or drops compared with the annual average sea level. The global sea level has risen by 10 to 20 centimeters in the recent years and shows an accelerating rise tendency. However, actual changes of the sea level in a given area of the world are also conditioned by local vertical land movements: slow crustal movements and local ground subsidence. As a slow-onset marine disaster, the cumulative effects of long-term sea level rise will aggravate the disaster severity of storm tides, coastal erosion, sea water intrusion, soil salinization, salt water intrusion and other marine disasters and cause related disasters.

The most direct influence of sea level rise on the coastal ecosystem of China is the loss of a large area of coastal salt marshes and tropical coral reefs, mangroves and other habitats. Moreover, in this long-term trend of sea level rise, important economically developed regions of eastern China will gradually become the coastal lowlands, with increasingly smaller development space and being more susceptible to natural disasters from ocean and land.

The past hundred years has seen an obvious tendency of global warming and a marked growth of the sea level rise rate. In the 20th century, the average of the global sea level rise rates was 1.7±0.5 mm/year; the average rise rate of the global sea level from 1961 to 2003 was from 1.8±0.5 mm/year; the average rate of sea level rise from 1993 to 2003 was 3.1±0.7 mm/year. In recent years, the sea level in China has risen by 2.6 mm/year, higher than the global average. In 2010, the Bohai Sea, the Yellow Sea, the East China Sea and the South China Sea level rose by 64 mm, 75 mm, 66 mm and 64 mm, representing an average rise rate of 2.5 mm/year, 2.8 mm/year, 2.8 mm/year and 2.5 mm/year respectively, compared with common years.

> In 2012, China's coastal sea level reached the highest level since 1980, with an increase of 122 mm over an average year. Coastal air temperature and sea temperature rose by 0.4°C and 0.3°C respectively than that of an average year; and the air pressure dropped by 1.2 hPA than that of an average year. Compared with an average year, the rise margins of the Yellow Sea, Bohai Sea, East China Sea, South China Sea all exceeded 100 mm, among which the Bohai Sea underwent the mildest sea level rise of 31 mm. Under the influence of various factors, such as climate changes and accumulative effects of sea level rise, coastal areas in Liaodong, Shandong, Jiangsu and some other provinces were seriously affected by shore erosion, seawater invasion and soil salinization. In 2012, the high sea level intensified the influence of storm surge on some coastal areas like Jiangsu, Zhejiang and Guangdong, causing harm to production and life of local people and economic and social development.

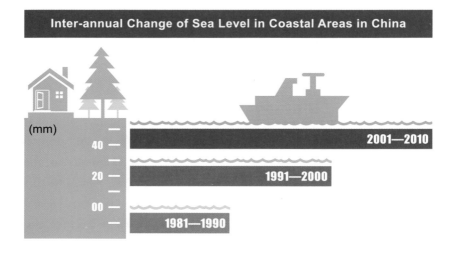

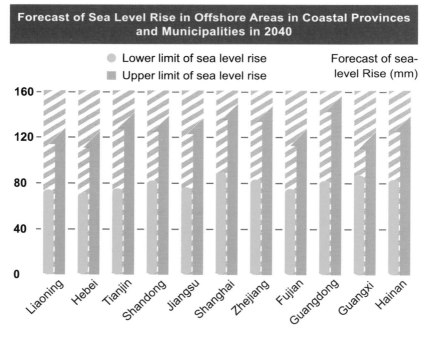

In the recent 30 years, China's decadal sea-level has also risen obviously in coastal waters. Since 2001, China's coastal sea level as a whole has been at a historic high point. The average sea level from 2001 to 2010

rose by 25 mm compared with the sea level from 1991 to 2000, and by 55 mm compared with the sea level from 1981 to 1990.

In future, low-lying coastal areas in China will suffer from direct threat from sea level rise. Relative sea level rise will not only directly inundate those low-lying areas, but also keep weakening disaster resistance abilities of waterproof projects in the coastal areas. In some coastal sections of the Bohai Rim, Yellow River Delta, Yangtze River Delta and Pearl River Delta, ground subsidence will be quite serious, and sea level rise will be very obvious.

> Since 2011, China has carried out nationwide risk investigation of large-scale projects and special assessment work, of which sea level rise risk assessment and zoning is one of the main work contents. In 2012, China compiled the "Technical Guidelines for Risk Assessment and Zoning of Sea Level Risk", which classifies the risk level of coastal regions nationwide,and provides policy support for marine economic development layout, exploitation and utilization planning of marine resources and construction of defense facilities for coastal large engineering projects of coastal areas.

According to the forecast, the sea level of China's offshore areas will have risen by 70 to 140 mm by 2040, and Tianjin, Shandong Province, Shanghai, Zhejiang Province and Guangdong Province will see their sea level rise get to the highest level, which may exceed 130 mm.

Coastal Erosion

Coastal erosion refers to coastline retreat, and shoal erosion and other phenomena as a result of the sediment supply less than incoming sediment under the marine force produced by natural factors, human factors or both. Natural factors include river sediment changes, sea level rise, storm surge, and sea wave attack etc., while human factors include coastal sand excavation, sediment interception in river water conservancy, unreasonable coastal engineering, land reclamation by sea encircling and aquaculture. All these will lead to erosion of local or remote coasts. Coastal erosion may directly give rise to erosion and inundation of surrounding land, and destruction of embankment dams, ports, buildings and structures, coastal defenses, protective forests, coastal road network settings, bathing beaches and other tourist facilities. Indirect disasters mainly refer to changes of the ecological and natural environment of the coastal zones.

Coastal erosion is a general disastrous geological phenomenon in global coasts. Affected by global warming and sea level rise, erosion of coasts worldwide has shown an aggravating trend in recent decades. Due to the improper development, China's coastline loss increases year by year. According to incomplete statistics, the total length of the shoreline

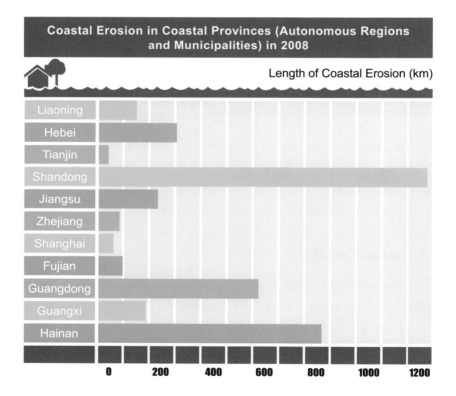

erosion in 2008 was up to 3,708 kilometers, with sandy shoreline erosion of 2,469 kilometers, accounting for 53% of the total sandy shoreline, and muddy shoreline erosion of 1,239 kilometers, accounting for 14% of the total muddy shoreline. The most serious sandy shoreline erosion takes place along the coast of Liaoning Province, Hebei Province, Shandong Province, Guangdong Province, Guangxi Province and Hainan Province, while the most serious muddy shoreline erosion occur along the coast of Hebei Province, Tianjin, Jiangsu Province and Shanghai. Shandong's shoreline that has been eroded has exceeded 1,200 kilometers, followed by

Coastal Erosion along Key Monitored Coasts in 2012

Province (Autonomous Region and Municipality)	Key Coast	Type of Eroded Coast	Length of Monitored Coast (km)	Length of Eroded Coast (km)	Average Erosion Rate (km)
Liaoning	Suizhong	Sandy	112.0	58.1	1.9
	Gaizhou	Sandy	21.8	18.5	2.9
Hebei	Luan River estuary to Dai River estuary	Sandy	105.4	5.1	11.0
Jiangsu	Lianyungang to Sheyang estuary	Silt muddy	267.2	90.2	10.4
Shanghai	Chongming Dongtan	Silt muddy	48.0	3.4	22.1
Guangdong	Chikan Village of Leizhou City	Sandy	0.8	0.2	3.0
Hainan	Zhenhai Village of Haikou City	Sandy	1.5	0.9	7.0

Guangdong Province and Hainan Province with more than 600 kilometers.

China's sandy and muddy shore shoreline suffers from severe erosion, and such erosion in some areas shows an accelerating trend. Coastal erosion will lead to loss of land, destruction of housing, roads, coastal engineering, tourism facilities and breeding areas and finally great economical loss to the coastal areas.

Seawater Intrusion

> Seawater intrusion refers to encroachment of sea water or highly mineralized saline ground water with hydraulic connection with sea water in the landward direction along the water bearing strata, as an immediate result of destruction of the balance between freshwater and seawater at the water bearing strata of the waterfront area caused by changes to hydrodynamic conditions of groundwater in the waterfront area for the reason of natural or human factors.

Seawater intrusion can mainly be attributed to sea level rise, land subsidence and storm surge. Additionally, excessive exploitation of groundwater, unreasonable oil and gas exploitation and large-scale city construction will lead to ground subsidence and cause seawater intrusion. Seawater intrusion worsens ecological environment, creates drinking water problems for humans and livestock, limits industrial development, causes land salinization and substantially reduces agricultural yield.

Seawater intrusion is more severe in the coastal plain areas of the Bohai Sea, mainly Panjin City in Liaoning Province, Qinhuangdao City, Tangshan City, and Cangzhou City in Hebei Province, Binzhou City and Weifang City in Shandong Province, where seawater intrusion is generally 10 to 30 kilometers to the coast. Seawater intrusion in Liadong Bay and Laizhou Bay features a large scale and a high salinization level. Seawater intrusion in coastal areas in the north and both sides of the Liaodong

Bay covers more than 4,000 square kilometers, including 1,500 square kilometers of severe intrusion. In Panjin City of the Liaodong Bay, Seawater intrusion extends as far as 68 kilometers. In the Laizhou Bay, the seawater intrusion area reaches 2,500 square kilometers, including 1,000 square kilometers of severe intrusion. The farthest seawater intrusion distance is at the southern part of the Laizhou Bay and reaches up to 45 kilometers.

Sea water encroachment has a smaller influence on the coasts of the Yellow Sea, the East China Sea and the South China Sea. In Yancheng City in Jiangsu Province and Taizhou City in Zhejiang Province, sea water intrusion along the coasts extends farther, whilst sea water intrusion in other areas is usually within 5 km away from the shore.

The Chinese government tried its best to forecast the sea ice disaster and release information during the disaster. According to "Emergency Response to Storm Surge, Tsunami and Sea Ice Disasters", it started the emergency response and organized three provinces and one city around the Bohai Sea to carry out effective emergency management and defense of the sea ice disaster. It paid close attention to the developments of the ice sea and used advanced devices like planes, survey boats, radars, satellites, etc. for emergency observation of the sea ice disaster, which provided accurate, timely and comprehensive information of sea ice monitoring , early warning and disaster conditions. It also adopted various ways to deliver the disaster information promptly and monitored and tracked fishing boats. The government deployed icebreakers emergently to carry out comprehensive and stratified de-icing and icebreaking work, and fixed saving devices like port facilities on time to mitigate the loss from the sea ice disaster.

Prevention of Marine Natural Disasters

The Third-level National and Sea Area Forecast Service System of the Coastal Provinces

In sea disaster prevention and reduction, China always thinks highly of the formulation of "Emergency Response to Storm Surge, Tsunami and Sea Ice Disasters" for which it has organized marine departments at all levels and formulated a series of emergency and execution plans, continuously expanding the coverage of such plans. In doing so, the national sea disaster emergency system has taken its initial shape. With the rapid development of marine economy, Chinese coastal areas are attaching greater and greater importance to sea disaster influences, and have been gradually carrying out assessments of national sea disaster risks. Coastal marine departments in various provinces have investigated and assessed the effects of sea level changes, determined the warning water level, drawn up and distributed communiqués on sea level and sea disasters, and constructed demonstration areas for marine disaster prevention and reduction, providing basic data and decision-making basis for effective reduction of disaster loss in various coastal areas.

Currently, China has a national forecast center and three sea area forecast centers, and 11 coastal provinces (regions, cities) have set up sea observation and forecast institutions. The sea forecast system has been

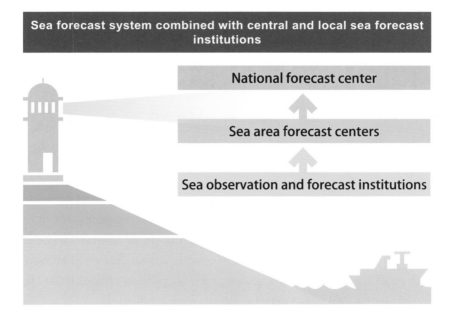

initially formed, which consists of national ocean environment forecast centers, sea area forecast centers and local ocean forecast institutions at all levels. The forecast of sea waves, sea temperature, ocean current and key port tides in the four large sea areas of the Bohai Sea, the Yellow Sea, the East China Sea and the South China Sea, together with the forecast of sea temperature, ocean current and sea surface winds of the northwestern Pacific Ocean and the global ocean wave, is made public every day through radio, television, newspapers, internet, SMS (short message service) and other channels. The sea ice forecast of the Bohai Sea and the north Yellow Sea is released every winter. The precaution alarm of various sea disasters is released on time during emergency response to storm surge, waves, tsunamis and ice disasters. Currently, marine disaster prevention and mitigation have achieved comprehensive improvements in terms of the management system, ability construction and operation mechanism. The duties of emergency management are being performed comprehensively and notable socioeconomic benefits have been achieved.

Coping With and Adjusting to Climate Changes in a Positive Manner

As a contracting party of the "United Nations Framework Convention on Climate Change" and the "Kyoto Protocol", China always attaches great importance to addressing climate changes. In 2007, Chinese government published "China's National Climate Change Program", which specifies coastal zones and areas as one of the key areas to adapt to the climate change. According to the "Twelfth Five-year Guideline" released in 2010, marine areas are explicitly required to accelerate their formulation and implementation of important laws and regulations and policies related to adaption to climate changes, and coastal areas are to actively carry out restoration of the ecosystems on islands and in coastal zones and offshore areas and strengthen coastal greening and vegetation rehabilitation, to effectively respond to and mitigate climate change effects.

China has introduced a series of special plans like "Advice on Work of Marine Areas in Response to Climate Change", "The Twelfth Five-Year Guideline for Oceanographic Technology Development", "The Twelfth Five-Year Plan for Marine Renewable Energy", "The Twelfth Five-Year Special Plan for Technological and Scientific Development in Response to Climate Change", etc., with specific advice on the way to strengthen marine climate observation, scientific researches, international communication and cooperation, etc.

China actively takes part in international cooperation plans like Global Ocean Observing System (GOOS), Global Sea Level Observing System (GLOOS), Climate Variability and Predictability (CLIVAR), Global Regular Assessment of Marine Environment (GRAME), etc. and plays an important role in the construction of Northeast Asian Region Global Ocean Observing System (NEARGOOS), Southeast Asia Global Ocean Observing System (SEAGOOS), and Indian Ocean Observing System (IndOOS). China has also signed the bilateral cooperation agreement on response to climate changes in marine areas with Indonesia, Malaysia, Thailand, New Zealand, Australia, etc. and set up the "Indonesia-China Center for Ocean and Climate", and cooperates with Italy to carry out the "Ecosystem Adaptation to Climate Change in Coastal Areas of China".

Setting up an All-day, All-round Marine Disaster Observation System

In 2013, Jiang Xingwei, Director of the National Marine Environmental Forecasting Center director under the State Oceanic Administration (SOA), said in an interview, "At present, China has set up an all-day, all-round marine disaster observation system, which can make prompt and accurate forecast to minimize disaster losses.

According to report on People's Daily Online, Jiang Xingwei believes that disaster mitigation must be based on forecast, which, in turn, necessitates intensified efforts on observation. At the moment, various ways are employed in China's marine disaster observation: first, all-weather monitoring through shore stations, buoys, ships and other means; second, frequent exchange of meteorological and hydrographic data with international counterparts; third, monitoring through sea satellites launched by China. He explained with an example, "When the typhoon 'Rananim' was in the East China Sea from 8:00 AM on 11st to 8:00 PM on 13th, August, the No. 9 ocean observation buoy deployed by the SOA recorded the sea wave and meteorological data accurately and completely throughout this typhoon process and observed the largest wave height of 13.2 meters, providing valuable data to marine disaster forecast."

Jiang Xingwei also said, "Currently, China has set up the marine disaster warning mechanism with tsunami warning included and the marine disaster prevention and mitigation emergency response system. At 4 p.m. on September 14th of last year, the National Marine Forecast Observatory and Tianjin Marine Environmental Monitoring and Forecasting

Sea water detection buoys of 2008 Olympic sailing boats in Qingdao

Center issued in time warning of the storm surge and rough sea. Tianjin municipal government started the marine disaster prevention and mitigation emergency response system on time and effectively organized disaster prevention and mitigation, minimizing the disaster losses."

According to Jiang Xingwei, in order to realize the professional operation of numerical wave prediction in Chinese seas and its adjacent waters, now Chinese marine science and technology workers are developing the numerical wave forecast model in deep and shallow waters and the numerical forecasting model for wind field at the sea surface and of typhoon applicable to limited regions covered by marine environment forecasts. They are also developing high resolution storm surge - near shore wave coupling numerical forecast model, typical regional storm tide floodplain numerical forecast model, El Nino regional and global numerical prediction model, etc. With the development of science and technology, the capacity of Chinese Marine disaster prevention and mitigation will continuously improve.

CHINA'S MARINE CONSERVATION AND DEVELOPMENT

THE AZURE OCEAN
—CHINA'S MARINE ENVIRONMENT CONDITIONS IN COASTAL WATERS

On July 7th, 2011, preliminary operation contest of North Sea of the 1st Chinese marine environment monitoring professional technology competition was held on boat "Red Sun 08" in Jiaozhou Bay, Qingdao.

The overall quality of Chinese marine environment has been deteriorating since the end of the 1970s, and pollution incidents damaging events happen from time to time. In recent years, offshore environmental problems have become prominent, whilst the environment quality in other sea areas is better. Offshore areas occupy about 5% of sea areas under Chinese jurisdiction. The main environmental problems include poor water quality, ecosystem degradation, red tide disasters, local seawater intrusion, soil salinization, and coastal erosion and emergency incidents like oil spills at sea.

Offshore Water Quality

The total area of offshore polluted waters in China in recent 10 years remains about 140,000 Sq km, with heavier pollution. In 2012, it reached 170,000 Sq km, over 40% of which was heavily polluted waters.

Heavily polluted waters is distributed mainly over medium and large estuaries, bays and offshore areas of some major cities, among which offshore areas of the Liaodong Bay, Bohai Bay, Laizhou Bay, coastal areas in Jiangsu Province, Yangtze River estuary, Hangzhou Bay and Pearl River estuary are seriously polluted for a long time with extremely poor water quality. Main pollutants in waters are inorganic nitrogen, active phosphate and petroleum substances

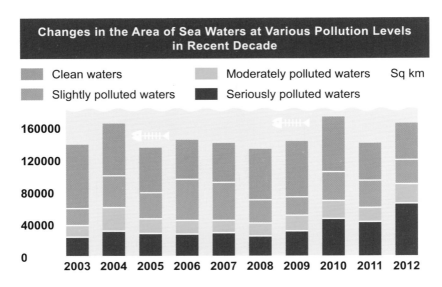

Typical Ecosystem Health

Monitoring results of China's offshore ecologically fragile areas and sensitive areas conducted by SOA shows that ecosystems in China's main bays, estuaries and coastal wetlands are all in a sub-healthy or unhealthy state, and those in Hangzhou Bay and Jinzhou Bay has been in an unhealthy state year in year out. Ecosystems in the east coast seaweed beds of Hainan Island, coral reefs and mangroves in Beihai, Guangxi, and coral reefs in Leizhou Peninsula maintain in a healthy state. Overall ecosystems of offshore waters are in a fragile state and the trend of ecological deterioration has not been mitigated.

The ponds for shrimp farming reclaimed by local fishermen are approaching mangroves in Jiangshan Peninsula, Fangchenggang City, Guangxi.

Unhealthy ecosystems mainly represents in eutrophication and nutrient imbalances, abnormal biocenoses structure, severely degraded estuarine spawning grounds and some gradually disappeared spawning grounds. The main influencing factors are land-based pollutants discharged into the sea, encroachment of marine environment by reclamation and over-exploitation of biological resources.

Red Tides and Green Tides

Red tides are a harmful ecological anomaly where tiny species of phytoplankton, protozoa or bacteria accumulate rapidly in the seawater under certain environmental conditions, resulting in discoloration of the surface water. Different causes to red tides, and different types and quantities of red tides can lead to different colors of the seawater, mainly including red or brick red, green, yellow, and brown. It is worth noting that when some of the red tide organisms (such as gonyaulax, gymnodinium and pyrocystis) cause red tides, they may not cause the water to present any particular color.

Red tides are mainly formed by dinoflagellates and diatoms with a small amount of protozoa and bacteria. Some red tide organisms pose no threat to humans, but they can produce toxins which harm fish and other marine organisms and damage marine ecosystems. However, other red tide organisms, though nontoxic, can cause blockage or mechanical damage to other marine organisms, and may also cause fish suffocation due to large consumption at the time of their death. Red tides can also cause enormous damage to the marine environment, such as the degree of acidity and the intensity of illumination in seawater, resulting in serious harm to the marine ecosystems.

On May 9th, 2014, red tides flooded into the Xiamen Port at breakneck speed in Xiamen City, Fujian Province. In less than half an hour, the Lujiang had turned red and the Gulang Island was in dark red waters.

With the rapid development of modern industrial and agricultural production, coastal populations increase and a large number of agricultural and industrial wastewater and domestic sewage are discharged into the sea, a considerable part of which are directly discharged into the sea before treatment, causing even worse eutrophication in coastal waters and estuaries. Besides, the increased level of coastal development and the expansion of mariculture have also brought pollution problems to the marine environment and mariculture; the development of the shipping industry has led to the introduction of alien species of harmful red tides; the global climate change has also given rise to frequent occurrence of red tides.

Red Tides Occurred in China during 2001-2012

Year	Number of Occurrences	Distribution Area (Sq km)	Direct Economic Loss (RMB 100 million)
2001	77	15,000	10
2002	79	10,000	0.23
2003	119	14,550	0.43
2004	96	26,630	0.01
2005	82	27,070	0.69
2006	93	19,840	-
2007	82	11,610	0.06
2008	68	13,738	0.02
2009	68	14,102	0.65
2010	69	10,892	2.06
2011	55	6,076	0.03
2012	73	-	20.15
Total	833	169,658	34.33

Resource: State Oceanic Administration, *China Oceanic Disasters Bulletin (2001-2012)*

Red tides have become worldwide public hazards, which occur frequently in over 30 countries and regions including the U.S., Japan and Canada. Over the past 40 years, China has witnessed continued deterioration of environmental quality in offshore areas of China seas, and frequent occurrences of red tides, with the frequency increasing by three times every 10 years. Since 2001, China has been in a period of frequent occurrence of red tides, with the year of 2003 being a period with the most frequent outbreak of red tides, totaling 119 times; after 2003, the incidence rate of red tides are on a downward trend. Red tides accouted for a much large portion of the marine areas in 2004 and 2005, outnumbering 25,000 sqare kilometers in respectively, while it was significantly reduced in 2006 in general. In 2012, the direct economic loss caused by red tides reached the highest level with the total loss of RMB 2.015 billion.

With China's economic development in coastal areas, the eutrophication in coastal areas becomes severe, and red tides have been found in each sea area, becoming one of the worst marine disasters. The high occurrence area of red tides is the East China Sea, mainly along Zhejiang coast. The Bohai Sea is also a red tide-prone area. In recent years, the frequency of occurrence of red tides in the Bohai Sea has increased, with the affected area spreading from part of inshore waters to the whole Bohai Sea area; the time of occurrence of red tides has extended to last from April to November. From 2001 to 2012, serious red tides occurred in China, including 19 red tides caused by dinoflagellates, accounting for 48.7% of all serious red tides in the same period containing once in the Bohai Sea and 18 times in the East China Sea.

Green Tides: Explosion and Accumulation of Large-scale Green Algae

Green tide is a harmful ecological phenomenon where some macrophytic species of green algae (such as enteromorpha prolifera) proliferate and accumulate rapidly in the water under certain environmental conditions, resulting in discoloration of the surface water. Like red tides, green tides are also marine ecological disasters. During the proliferation, the macrophytic green algae cover large sea areas and consume large amounts of oxygen in seawater, causing suffocation to other marine organisms. A large number of algae are brought by tide water and accumulate in the coastal areas, seriously affecting coastal landscapes and causing air pollution.

In 2004, a green tide occurred in the Yalong Bay in Sanya City of

> Enteromorpha prolifera
> In plant taxonomy, enteromorpha prolifera belongs to green algae. There are about 40 species of enteromorpha prolifera in the world, including about 11 sepcies in China. The common species of enteromorpha prolifera in China include enteromorpha linza, enteromorpha compressa and enteromorpha clathrata. The body of enteromorpha prolifera is in grass green, in tubular shape, and covered with membrane. It grows quickly, with outstanding main branches and slender branches, up to one meter high. Its root is attached to rocks via its holdfast, and grows on gravel on beaches in the intertidal zone. Due to global climate change and eutrophication and for other reasons, green tides caused by macroalgae enteromorpha prolifera often occurr in the sea. A large amount of enteromorpha prolifera floats and gathers on the shore, blocking waterways, causing damage to the marine ecosystems, and seriously threatening fishery and tourism development in coastal areas. Although enteromorpha prolifera plants are very slender, with a filamentous shape in green, this size is enough to let people call it "macroalgae", because it is constituted by mutiple cells. Therefore, compared with those single-cell algae with a diameter ranging from only a few microns to several hundred microns, enteromorpha prolifera can totally be regarded as a "huge monster."

Hainan Province, which marked the earliest record of green tides in China. From 2005 to 2007, green tides occurred in Shandong Province and Hainan Province respectively, causing a low degree of hazard. The green tide that occurred on the eve of the opening of the Olympic Games in 2008 was China's worst disaster, causing direct economic losses of RMB 1.322 billion. The green tide in 2009 affected the largest sea area of up to 58,000 square kilometers, causing direct economic losses of RMB 641 million. In 2012, the distribution area and the coverage area of enteromorpha prolifera both signaled the lowest in the past five years, namely 19,400 square kilometers and 261 square kilometers respectively.

Enteromorpha prolifera is usually formed in the central and southern

Yellow Sea waters in May and June each year and enters the coastal waters of the southern Shandong Province in July and August. In Qingdao City only, enteromorpha prolifera that has been salvaged and cleaned reached 60,000 tons. In recent years, enteromorpha prolifera disasters continue to occur and are featured by a larger occurrence area and a longer time of duration. The influx of enteromorpha prolifera into coastal waters of the southern Shandong Province has serious impacts on fisheries, aquaculture, marine environment, landscapes and ecosystem services.

On May 30th, 2008, large-scale enteromorpha prolifera appeared in the central Yellow Sea waters, floating rapidly towards Qingdao with the help of ocean currents and wind. During drifting, the enteromorpha prolifera grew and reproduced rapidly. On June 12th, it invaded Qingdao coast and Olympic Sailing field. On June 28th, the maximum distribution area of enteromorpha prolifera on the sea reached up to 24,000 square kilometers, of which 16 square kilometers were distributed in the Olympic Sailing field with an area of 50 square kilometers, causing a severe situation to the upcoming Olympic sailing events. This enteromorpha prolifera was the most serious natural disaster ever faced by Qiangdao over the years in terms of the accumulation scale, duration and difficulty in treatment.

Once the floating enteromorpha prolifera was monitored in the central Yellow Sea, the Qingdao municipal government immediately launched an emergency surveillance and monitoring mechanism in accordance with the "Contingency Plans for Green Tide Disasters", and monitored and tracked the development and drifting of the enteromorpha prolifera. On June 12th when the enteromorpha prolifera invaded the surrounding waters of the Dagong Island, the local government quickly arranged vessels to carry out interception and salvage, launched an emergency response plan for Class I according to the development and disposal of enteromorpha prolifera, rapidly transmitted information, mobilized fishermen, gathered fishing vessels, and deployed supplies, saving time for effective disposal of the enteromorpha prolifera.

Basic Situations of Green Tide Disasters

Year	Affected Area	Distribution Area (Sq km)	Coverage Area (Sq km)	Species of Green Tide
2004	Sanya of Hainan	-	-	Monostroma nitidum, enteromorpha prolifera
2005	Yantai of Shandong	-	-	Cladophora, enteromorpha prolifera
2006	Yantai of Shandong	-	-	Cladophora, enteromorpha prolifera
2007	Sanya and Qionghai of Hainan Qingdao of Shandong	-	-	Enteromorpha prolifera
2008	Qingdao of Shandong	25,000	650	Enteromorpha prolifera
2009	Coastal waters in southern Shandong	58,000	2100	Enteromorpha prolifera
2010	Rizhao, Qingdao, Weihai and Yantai of Shandong	29,800	650	Enteromorpha prolifera
2011	Rizhao, Qingdao, Weihai and Yantai of Shandong	26,400	560	Enteromorpha prolifera
2012	Coastal waters in southern Shandong	19,400	261	Enteromorpha prolifera

After the occurrence of the enteromorpha prolifera disaster, the Qingdao municipal government immediately held a press conference

On July 5th, 2014, 2 boats moored at sea covered with enteromorpha in Qingdao, Shandong Province.

to introduce to the public the characteristics, origin, mechanism of the occurrence, and current disposal of enteromorpha prolifera. They actively spread to all sectors of community the information such as: enteromorpha prolifera is nontoxic and has no direct harm to the environment; enteromorpha prolifera flowed in from open seas and had no direct relationship with the local marine environment; the government was organizing people to carry out an orderly disposal work to ensure the holding of the Olympic sailing events as scheduled.

The Qingdao municipal government quickly organized a removal team, responsible for cleanup and transport of the enteromorpha prolifera on shore. The departments of marine, maritime, port and waterway, and weather, as well as governments in regions and cities along the coast organized an offshore salvage fleet composed of more than 1,500 fishing boats and more than 8,000 fishermen. The fleet adopted the approach of "interception", "scooping" and "cleanup" and used large trawlers, medium ships with attack nets and small boats with hand nets to carry out large-scale salvage operations at sea. At the same time, the marine department of Shandong Province organized a joint fleet consisting of 920 large fishing vessels and more than 10,000 fishermen from coastal cities of Weihai, Yantai, Rizhao, Weifang, Dongying and Binzhou to perform interception and salvage in open seas. Jinan Military Region and the transport agency of Shandong Province were also engaged into the removal and transport work.

After the disposal of the enteromorpha prolifera, the Qingdao City actively dealt with the aftermath. For fishermen participating in the salvage, Qingdao City gave them appropriate subsidies in accordance with the actual fuel consumption. For dedicated facilities used for disposal of the enteromorpha prolifera, the city arranged relevant departments to handle them properly in accordance with their actual situations. The facilities might be stored for later use, put into business class, retired, or used for other purposes. The government encouraged enterprises to participate in innovating technologies to realize comprehensive utilization of enteromorpha prolifera. The enterprises developed a series of products including food additives, seaweed fluid and seaweed fertilizers, and innovated technologies and equipment for rapid drying of large scale enteromorpha prolifera. The government also organized a delegation to conduct investigation in countries frequently affected by green tide disasters, learn disposal experience, and introduce advanced technologies and equipment for salvage at sea, collection on shore, and rapid dehydration of enteromorpha prolifera. Based on its experience in disposal of enteromorpha prolifera, Qingdao City once again organized parties concerned to adjust and improve the original emergency plans for enteromorpha prolifera, which laid a solid foundation for scientific disposal of large-scale floating algae.

"Dead Zones" in Offshore Waters

A "dead zone" refers to a low-oxygen and oxygen-deficit zone in offshore waters. The land-sourced pollution and the global climate change are the two major reasons why a "dead zone" is formed. A large number of pollutants containing nitrogen and phosphorus are discharged into the coastal waters, resulting in the booming of algae and other plants. When these plants die and sink to the bottom of the sea, the decomposition process requires consumption of lots of dissolved oxygen. As the sea water temperature rises, the dissolved oxygen in seawater will be less and less. At the same time, the increase in oxygen consumption of aquatic organisms

Since the early June 2014, enteromorpha "Forward" reached many sea areas in Lianyungang, Jiangsu Province, some of which were pushed to shore, and the staff was cleaned up them.

may, to a certain extent, exacerbate ocean oxygen deficiency. Since oxygen deficiency leads to death of fish, crustaceans and other animals and plants, the entire waters becomes a "dead zone" without living creatures.

In the past 50 years, the area of "dead zones" has continued to expand. There are already 400 such "dead zones" in the world, covering a total area of more than 240,000 square kilometers. Among them, the Gulf of Mexico "dead zone" is caused by the enrichment of lots of agricultural fertilizers and other human-induced emissions containing rich nitrogen and phosphorus flowing from the Mississippi River into the Gulf of Mexico.

> **Dead Zones**
> According to the frequency and duration of anoxic events, "dead zones" can be divided into temporary "dead zones", cyclical "dead zones", seasonal "dead zones", and persistent "dead zones".
> Temporary "dead zones" refer to waters where the anoxia is occasional and the recurring cycle is less than one year.
> Cyclical "dead zones" refer to waters where anoxic events occur and last for several hours to several weeks each year.
> Seasonal "dead zones" refer to waters where anoxic events generally occur every summer or autumn.
> Persistent "dead zones" refer to waters where the duration of hypoxia is very long and even goes beyond a year.

In recent years, large-scale red tides have occurred in coastal waters in China, and the oxygen deficiency problem at the bottom of coastal waters near the Yangtze River Estuary, Pearl River Estuary and East China Sea has also been exacerbated. The Yangtze River Estuary is listed by the United Nations Environment Programme as a permanent "offshore dead zone" extremely difficult to recover, while the Pearl River Estuary and Zhejiang offshore area are listed as seasonal "offshore dead zones".

Oil Spills

Marine oil spills refer to incidents where oil leaks into the sea during oil production, refining, handling, storage, transport, use and disposal, causing damage to the marine aquatic organisms and ecological environment. Since the 1980s, with the rapid development of the maritime industry and the offshore oil industry, China seas have experienced frequent oil spills. According to statistics, during the period from 1973 to 2009, the coast of China experienced 2,821 oil spill accidents, once every 4-5 days. Oil spills have caused serious impacts and economic losses to the fishery industry, coastal tourism and marine environment, and are endangering human health.

In addition, with the increase in intensity of offshore oil and gas development, oil pollution incidents occur frequently and are been on the rise in recent years due to leakage of offshore oil and gas platforms and oil pipelines. In 2010 and 2011, five consecutive oil spill accidents occurred to China National Petroleum Corporation (CNPC) in Dalian's Xingang Harbor, hitting the local tourism and aquaculture. On June 4th and June 17th of 2011, two oil spill accidents hit Penglai "19-3" oilfield one after another, resulting in the flow of lots of crude oil and oil-based mud into the sea, causing serious pollution and damage to the marine environment of the Bohai Sea.

Penglai "19-3" oil spill accident was the first undersea oil spill accident in China and has caused huge pollution and damage to the marine environment of surrounding waters. Affected by the oil spill accident, a total area of 6,200 square kilometers of seawater was polluted (exceeding the sea water quality standards for Category I), of which 870 square kilometers of seawater was severely polluted (exceeding the sea water quality standards for Category IV). Penglai "19-3" oil spill accident has caused significant reduction in the variety and diversity of marine organisms in the polluted waters near the Bohai Sea and in northwest Bohai Sea, and severely affected the ecosystems and fishery production in the Bohai Sea. It was the most serious accident since China's commencement of marine resources development.

After the occurrence of the oil spill accident, the SOA immediately instructed ConocoPhillips to conduct investigation into the oil spill accident, and took the initiative to publish relevant information to the public in accordance with the requirements of the "Executive Program on Emergency

Penglai "19-3" oil spill site in Bohai Sea

Response to Oil Spill Accidents Occurring during Offshore Oil Exploration and Development". The SOA completed the blockade of the spilled oil source near Platform B prior to August 31st of 2011, and at the same time removed the undersea oil stains leaked from Platform C. In August 2011, the Chinese government established the Penglai "19-3" oil spill joint investigation team.

From August to November 2011, through site investigation and evidence collection, the Penglai "19-3" oil spill joint investigation team and SOA made a series of instructions, including instructing ConocoPhillips to stop re-injection, drilling, and oil and gas production operations; they concluded that ConocoPhillips had violated the overall development program during the production operations of Penglai "19-3" oilfield, as it failed to have relevant management and other systems in place, and had not taken necessary precautions after the appearance of obvious accident signs, thus causing a major marine oil spill accident.

To effectively safeguard the legitimate rights and interests of fishermen, and earnestly safeguard the interests of the national marine environment, SOA established a national leading group for marine ecological claims on August 30th. The group conducted scientific assessment according to standards such as the technical guidelines for assessment of ecological damage caused by marine oil spills, and estimated the amount of compensation for ecological damage caused by oil spills. Then, SOA raised a claim, on behalf of the country, to ConocoPhillips for ecological damage.

After administrative mediation, on January 25th, 2012, the Ministry of Agriculture, China National Offshore Oil Corporation (CNOOC), ConocoPhillips and relevant provincial government reached consensus on how to solve the issue of damages and compensation for fishermen due to Penglai "19-3" oilfield oil spill accident. ConocoPhillips contributed RMB 1 billion as compensation for damages to biological resources and natural fishery resources in the Bohai Sea near some counties of Hebei and Liaoning provinces; ConocoPhillips and CNOOC spent RMB 100 million and RMB 250 million respectively in establishing a marine environment and ecological protection fund to be used for restoration and conservation of natural fishery resources, as well as survey, monitoring, assessment and scientific research of fishery resources and the environment. In April 2012, ConocoPhillips and CNOOC agreed to pay RMB 1.683 billion as damages and compensation for losses of marine environmental capacity and marine ecosystem services caused by oil spills, for restoration of marine habitats, and for recovery of marine biological populations.

Land-sourced Pollutants

Land-sourced pollutants refer to pollutants discharged from the land into the sea. They are the main cause of pollution of China's marine

A few swans were leisurely finding food and resting in the Yanghe estuary of Qinhuangdao, Hebei. Local fishermen said that they hadn't seen swans here for almost ten years due to water pollution.

environment, accounting for over 80% of all pollutants into the sea. Land-sourced pollutants are discharged into the sea mainly through rivers and sewage outfalls, of which rivers are the main channels. Since most rivers in China are polluted, the amount of pollutants discharged into the sea through rivers remains high for years. In 2012, the total pollutants discharged into the sea through 193 major rivers included: 4.403 million tons of permanganate index, 623,000 tons of ammonia nitrogen, 61,000 tons of oil, 3.694 million tons of total nitrogen (TN), 316,000 tons of total phosphorus (TP).

Most sewage is discharged from cities into the sea without necessary treatment and more than half of the sewage outfalls discharge more sewage than allowed throughout the year. In 2012, the total quantity of sewage discharged from the 425 industrial pollution sources, domestic pollution sources and integrated sewage outfalls with daily emissions of over 100 cubic meters of sewage directly into the sea as shown by monitoring data, was about 5.60 billion tons. The total amount of each pollutant discharged

Total Amount of Each Pollutant Discharged Directly into the Sea through 193 Rivers in 2012 (Unit: Ten Thousand Tons)

	Permanganate Index	Ammonia Nitrogen	Petroleum	TN	TP
Bohai Sea	7.1	1.6	0.2	5	0.3
Yellow Sea	23.4	2.4	0.3	8.8	0.5
East China Sea	306.1	37.7	4.2	272.8	26.9
South China Sea	103.7	20.6	1.5	82.8	3.9
Total	440.3	62.3	6.2	369.4	31.6

Total Amount of Each Pollutant Discharged Directly into the Sea from 425 Pollution Sources in 2012				
	COD (Ten thousand tons)	Ammonia Nitrogen (Ten thousand tons)	Petroleum (Tons)	TP (Tons)
Bohai Sea	0.7	0.1	35.8	90.8
Yellow Sea	5.4	0.4	102.5	674.7
East China Sea	12.3	0.9	614.5	1,206.6
South China Sea	9.6	0.3	273.2	948.7
Total	28	1.7	1,026	2,920.8

into the sea was as follows: 280,000 tons of chemical oxygen demand (COD), 1,026.1 tons of petroleum, 17,000 tons of ammonia nitrogen, and 2,920.9 tons of total phosphorus.

The nutrient salt of high concentrations in the discharged sewage can lead to eutrophication of seawater and imbalance of nutrient salt. Serious eutrophication exists in about 70% of seawater, which can result in frequent occurrence of large-scale red tides in the central waters of Zhejiang Province, the waters off the Yangtze River Estuary, and the Bohai Sea waters. This can also result in simplified benthos and monotonous species in waters near the sewage outfalls, and even turn some waters near the sewage outfalls into undersea deserts without benthos.

Ecological Impact of Land Reclamation from the Sea

Coastal areas are the most developed areas in China, with large population, but less per capita land area than in inland areas. To solve the stiff problem of shortage of land resources, coastal provinces and cities have continued to take land reclamation from the sea as the major way to solve the problem of shortage of land resources.

Since 1949, China's coastal areas have experienced four waves of land reclamation from the sea. To meet the requirements of urban construction, port building and industrial construction, China launched a new round of land reclamation from the sea from the 1990s. From 1990 to 2008, the total area of land reclaimed from the sea increased from more than 8,000 square kilometers to 13,000 square kilometers. In recent years, the land reclamation from the sea has presented the development momentum of high speed, large area and wide range.

Land reclamation from the sea has changed the natural properties of sea areas, seriously affecting the offshore marine environment and the health of coastal marine ecosystems. On the surface, it is difficult for people to see and feel in a short time the more serious impacts on the environment caused by land reclamation, but the facts that can be seen for now are enough to prove the destructive effects of this practice.

According to a special investigation into islands and coastal areas lasting six years and published in 2011, in the past 20 years, a total of more than 700 islands disappeared, including over 200 islands in Zhejiang Province, over 300 in Guangdong Province, 48 in Liaoning Province, 60 in Hebei Province, and 83 in Fujian Province.

The land reclamation from the sea has artificially changed the locations of China's coastlines, which used to be in the equilibrium state formed after millions of years of interaction between ocean and land, and from which the wetlands and organisms near the coastlines have also benefited. Once the coastlines are moved artificially, this balance will be broken. The disappearance of islands as barriers will surely affect the status of coastal wetlands. According to statistics, the total area of China's coastal wetlands is 69,300 square kilometers, but in recent years, the area has decreased by more than 6,500 square kilometers compared with that in 1975.

The significant reduction in the area of coastal wetlands can even lead to red tides, floods and tsunamis. It can also reduce biodiversity and fishery resources, which is not conducive to flood discharge and can cause land subsidence in urban areas.

20% of the land area of the Netherlands is formed by land reclamation from the sea. From 1950 to 1985, the Netherlands lost 55% of its wetlands. The loss of wetlands in the Netherlands has caused many environmental problems in terms of pollutant degradation and climate regulation, such as coastal pollution and birds decline. In 1990, the Ministry of Agriculture of the Netherlands formulated the "Policy Plan for Nature", which specified that it would take 30 years to restore the country's "nature." According to the plan, some dikes on both sides of the Westerschelde waterway in the south of the Netherlands would be torn down, and the 300-hectare "polder"—the land reclaimed from the sea—would be submerged by the sea and reverted to a wetland for bird habitat.

It is estimated that in the past 100 years, Japan has acquired a total of 120,000 square kilometers of land from the sea. About 1/3 area of the coastal cities is obtained through such reclamation, and 95% of its natural coastlines become artificial shorelines. Such large scale land reclamation has caused damage to the environment, reduced tidal influx, and weakened self-purification capacity of seawater, resulting in deterioration of seawater quality and degradation of marine biological resources; the decline of natural wetlands has led to rapid decline of biodiversity along coastlines.

Recognizing the problem, the Japanese government began to examine the reclamation construction, and invested heavily every year in the establishment of a special "renewable grant project", hoping to find some way to restore the ecological environment. At present, the total area of land reclaimed from the sea is less than 1/4 of that in 1975 and the annual reclamation area is only about 5 square kilometers.

CHINA'S MARINE CONSERVATION AND DEVELOPMENT

PROTECT THE SOURCE OF LIFE
—ECOLOGICAL AND ENVIRONMENTAL PROTECTION OF CHINA SEAS

A sea is like a poem, alive and active; like a painting, rich in implications; like a symphony, vast and strong… It is sometimes calm, like a mirror; sometimes rage, filled with roaring waves. In my mind, a sea is a blue world and a cradle for many free lives, which makes it mysterious and charming!

However, a sea nowadays has been changed. The seawater is no longer clear. It becomes turbid and dirty. When the seawater rushes onto the beach, what are left on the beach are piles of garbage, mingled with dead fish and shrimps to emit foul smell. Why? All these can be attributed to our human beings. We often throw stink garbage into the sea when enjoying the beautiful scenery at the beach. People living by the sea lead the drain pipes of their toilets directly into the bottom of the sea. More appalling is that a large amount of wastewater is discharged from plants into the sea, dyeing the seawater into a strange color.

A person once did an experiment. He dripped a drop of ink into a pot of water which he regarded as a "sea". The ink slowly spread, became lighter and disappeared. The "sea" water was almost as clean as before; as he continued to drip ink into it, the color of the "sea" water got darker and finally turned into black. Six billion people on the planet produce a large amount of sewage and industrial pollution every day. If we discharge sewage into the sea like dripping ink into the pot of water, will there be life in the sea? What will the sea provide to us? I think there must be a lot of terrible things that have happened or are happening at the sea, by the sea and in the sea.

For the sake of our human health, and to ensure that our children and grandchildren can enjoy the beautiful and charming sea, let us take action now and love the ocean as we love our lives.

This is an essay written by a primary school student and released on the Internet in China on May 2nd, 2013. Reading this essay, all of us will be touched by it.

To protect the marine environment, China has enacted laws and administrative regulations concerning the environment, resources, and

Police were monitoring over the order and dynamic traffic over the port through high-powered telescope during their patrol in Penglai Port, Shangdong Province.

environmental and resource protection, to regulate the exploitation and utilization of marine resources and realize sustainable social and economic development.

The Establishment of Marine Environmental Protection System

The Marine Environmental Protection Law of the People's Republic of China was initially approved by NPC in 1982 and was put into effect in 1983. The current version is the outcome of two amendments made in 1999 and 2013 respectively, which is the fundamental law of the China's marine environmental protection. It has specified various systems concerning supervision and management of the marine environment, marine ecological protection, prevention and control of pollution damage to the marine environment caused by land-sourced pollutants, coastal construction projects, marine construction projects, waste dumping, shipping and related operating activities. For these systems, corresponding

At the closing meeting of the Sixth Session of the Standing Committee of the 12th National People's Congress held at the Great Hall on December 28th, 2013, amending seven laws including the Marine Environment Protection Law of the People's Republic of China were adopted.

On December 22th, 2009, the Hainan CMS Branch began to carry out joint marine, land and airplace law enforcement in seven cities and counties including Haikou City, Danzhou City and Sanya City.

supporting regulations have been formulated and promulgated, including the Regulation on the Administration and Prevention of Pollution Damage to the Marine Environment Caused by Marine Construction Projects, the Regulation of the People's Republic of China on the Administration and Prevention of Pollution Damage to the Marine Environment Caused by Coastal Construction Projects, and the Regulation of the People's Republic of China on the Administration and Prevention of Pollution Damage to the Marine Environment Caused by Land-Sourced Pollutants.

Chinese established a system of protecting the marine environment supervision and management system which is "unified supervision and management, division and classification of responsible ". The competent administrative department of environmental protection is responsible for the unified supervision and management of national environmental protection work. The competent administrative departments of marine, maritime, fishery and the military duties are responsible for their respective scope of authority and the work of marine environmental protection and management

The Marine Environmental Protection Law of the People's Republic of China has proposed to establish and implement a system to control the total amount of sewage in key waters and has required determining the main indicators to control the total quantity of pollutants discharged into the sea and control the quantity of major pollutants discharged into the sea. The law has also established a system to charge for the sewage discharged into the sea. According to the system, organizations and individuals who discharge pollutants directly into the sea must pay sewage charges on a regular basis according to national regulations, and those who dump waste into the sea must pay dumping fees according to national regulations.

Implacement of Marine Functional Zoning

The Law on the Administration of Sea Areas has established the marine functional zoning system to make overall arrangements of the use of the sea, which plays a leading role throughout the world. The main content of marine functional zonings is to scientifically determine the functions of the sea areas based on their natural attributes such as location, natural resources and natural environment, and make the overall arrangements of the use of the sea areas for relevant industries, to protect and improve the ecological environment and ensure rational use of the sea areas. The Functional Divisions of the Sea of the Whole Country (2011-2020) has proposed the guiding principle of "protecting in development, and developing in protection" for the use of the sea areas, selected a number of coast sections and sea areas with good basic conditions and low environmental sensitivity

In early winter, tens of thousands of sea birds are flying low in great flock or finding food which forms a charming view.

for focused construction and intensive use. At the same time, it has also put forward specified requirements to protect important habitats and typical ecosystems in estuaries, harbors, islands, and areas with concentrated distribution of mangroves and rare and endangered marine organisms, and drew red lines that development activities are not allowed to go beyond.

The Functional Divisions of the Sea of the Whole Country (2011-2020) implements control and management of the minimum retention of marine protected areas and determines that the total area of marine protected areas should reach more than 5% of the sea areas under the jurisdiction of China by 2020. In addition, in response to the excessive consumption of resources in China's sea areas and the prominent scarcity of coastal waters and coastlines, it has also put forward the objectives that the area of reserved areas in China's coastal waters should not be lower than 10% of the total area of China's coastal waters by 2020, and that the retention rate of natural coastlines on the mainland should not be lower than 35% by 2020.

According to the specific conditions in various sea areas, the

Functional Divisions of the Sea of the Whole Country (2011-2020) has proposed targeted sea area management measures. In the Bohai Sea, the most stringent reclamation management and control policies and the most stringent environmental protection policies will be implemented. The Yellow Sea coast with vast alluvial beaches, diverse marine ecosystems, and unique biotas is one of marine ecosystem areas that enjoy protection priority of the international community. It needs to be protected. In the East China Sea, the protection of bays, islands and surrounding waters will be strengthened, reclamation within the bays and reclamation to connect islands will be restricted, and the conservation of important fishery resources and aquatic genetic resources will be strengthened. In the Yangtze River Delta and the Zhoushan Islands waters, the system to control the total quantity of pollutants discharged into the sea will be implemented in order to improve the quality of the marine environment. In the South China Sea waters, the protection of marine resources will be strengthened, the scale of land reclamation from the northern coastal waters, especially the estuaries and bays, will be strictly controlled, and the construction of protected areas with islands and reefs as the protected targets will be accelerated.

Establishment of System of Island Protection

It requires the State Council and local people's governments at various levels in coastal areas to include the protection and rational exploitation of the islands into the national economic and social development plan, and take effective measures to strengthen the protection and management of the islands, and prevent the destruction of ecosystems on the island and in the

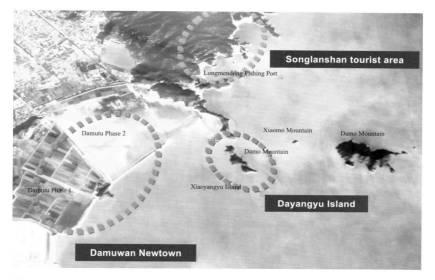

The Dayangyu Island in Xiangshan County, Zhejiang Province became China's first uninhabited island with the right to use for 50 years auctioned and "bought" by a company in Ningbo in November 2011.

surrounding waters. The law has established a number of important systems, including island protection planning system, island ecological protection system, uninhabited islands tenure and paid use system, special purpose island protection system and supervision and inspection system. The conservation plan of islands is the basis to take up the activities to protect and make use of the islands. The system of management authority and compensation for use in non-resident islands stipulates that the non-resident islands belong to the nation, the proprietary rights should be exercised by the the State Council. Protection system for special-purpose islands mainly specifies special protective measures on the islands which has the baseline points of the territorial sea, the islands used for national defense and the marine natural reserves.

Marine Environmental Protection Mechanism to Explore Sea Linkage

In recent years, environmental protection model concerning sea linkage is gradually paid attention to by relevant authorities in China. The Ministry of Environmental Protection, the State Oceanic Administration, Development and Reform Commission, the Ministry of water resources and other ministries have made efforts to promote the marine and coastal zone ecological environmental protection work under the "Both Land and

On December 12th, 2012, Guangdong, Hong Kong and Macao carried out joint law enforcement on marine environment protection for the first time in the sea areas around the Pearl Estuary.

Sea, River and Ocean" policy. At the same time, they carried it out at the regional level such as the Bohai Sea. Since 2001, a number of ministries have implemented two major marine environment protection plans in the Bohai Sea: Bohai Blue Sea Action Plan and General Plan of Environmental Protection of in the Bohai Sea. In 2001, it was approved and put forward by the State Council in Bohai Blue Sea Action Plan that we should balance land and sea, river and ocean, emphasis on the renovation ofland-based pollution, curb the deterioration of the sea environment, improve the quality of marine environment, work hard to enhance the marine ecosystem service function and to ensure the sustainable development of social economy in the Bohai Sea area.

In 2008, the State Council approved the Bohai Sea Environmental Protection Overall Plan (2008 - 2020) which was jointly implemented by the National Development and Reform Commission, Ministry of environmental protection, Urban and Rural Construction Ministry, the Ministry of Water Resources, the State Oceanic Administration and other departments. It was put forward that we should balance land and sea, river and ocean and it was required to strengthen the comprehensive control of pollution sources from ocean to rivers as well as from mouth to upper catchment area. The land-based sources of pollution control, river basin water resources, water environment comprehensive management and coastal protection should all be combined.

In 2012, China completed the "Marine Pollution Prevention Plan (2012—2015)". It aims at the improvement of offshore environmental quality and the protection of the marine ecosystem's health. It adheres to the "Coordinating Sea and Land, Balancing Rivers and Ocean" principle, analyses the situation of the prevention of coastal waterpollution, and defines basic tasks of five aspects and target tasks of the planning for forty key sea areas.

Maintaining Marine Fishing Resources in Multimodal Means

Chinese is the only country in the world whose aquaculture production exceeds its fishing. As a traditional fishery country, China used to rely on the natural fishery resources to meet the needs of the people of fish protein, and the increase in fishery production was mainly dependent on the increase in catches, which caused great pressure on the coastal and inland fishery resources. By 1978, fishing production accounted for 74% of the total output.

In 1985, China established the fishery development guideline "Mainly to Breed" to encourage fishermen to develop fishery production and thus rationally utilized and developed the resources of inland waters, shallow beaches and low-lying wastelands which were suitable for breeding but had been ignored for thousands of years. Since 1988, the amount of aquaculture production began to output the amount of fishing yield, and the structure of fishery production was optimized to some degree.

With the increasing need of aquatic products as well as the pressure on natural fishery resources it brought about, China gave priority to aquaculture development, conservation and reasonable utilization of fishery resources in 1997 which further optimized the structure of fishery industry. Since 1999, China started the implementation of the "Zero Growth" plan to stop marine fishing yield from growing. At present, marine fishing yield growth rate has witnessed a downward trend.

From June 1st to September 16th 2014, the East China Sea entered the fishing moratorium except fishing by fishing tackles, thousands of fishing boats all backed to the ports in Zhoushan City.

In the 1990s, the trends of fishing intensity exceeding resource reproducible ability and over-exploitation of fishery resources had shown in China. The facts of serious exhaustion of fisheries resources, significant reduction of the main economic fish resources, decline of marine fishery benefits, production halt of fishing boats, and drop of the income of fishermen had become serious economic and social problems in many places.

Summer Fishing Moratorium System

In 1995, China began to implement a comprehensive fishing moratorium system in the East China Sea, Yellow Sea and Bohai Sea.

According to the system, certain fishing operations cannot be carried out in certain period and in certain waters. Since the period of fishing moratorium determined by the system is in summers, the system is called summer fishing moratorium.

The implementation of summer fishing moratorium system in the East China Sea, the main economic fish resources such as belt fish have effectively been protected. Therefore, in 1999, the Ministry of Agriculture issued regulations, specifying to implement summer fishing moratorium in the South China Sea from the same year.

At present, all trawling, seining and hang trawling vessels, whether from mainland China, Hong Kong, Macao and Taiwan regions, or foreign countries, shall comply with the summer fishing moratorium system.

At 12:00 a.m. on September 16th, 2012, three and a half months' fishing-off season ended in the East Sea in China. Fishing started comprehensively. In the afternoon of 15th, a grand sea sacrifice ceremony was held in Shiqu Town, Xiangshan County, Ningbo City, Zhejiang Province.

Sea ceremony was held at the Shipudongmen fishing village in Xiangshan County, Ningbo.

Every year, fishermen will hold the sea sacrifice events one day before the fishing begins in order to praying for safety and bumper. The solemn and serene sea sacrifice ceremony, which has had an history of over 1,000 years, not only embodies fishermen's traditional thoughts of having reverence for nature and being grateful for oceans, but also embodies the modern idea that human and nature develop harmoniously, and protecting the oceans is exactly protecting ourselves.

Ecological Restoration for Seas and Islands

Coastlines are scarce marine space resources. But in recent years, China's coastal areas have witnessed problems of extensive utilization of coastlines and damage to coastal landscapes.

From 2009 on, China has been carrying out regulation and repair of coastal waters, especially those with damage to natural landscapes, degradation of ecological functions, diminished capacity for disaster prevention and mitigation, and low efficiency in utilization.

In China, the ecological restoration in fishing waters is a normal work in the coastal provinces and cities. The marine fishery management departments in coastal areas carry out multiplication release and occasionally release artificial reefs into the sea each year. Shandong and other provinces allocate special funds from the provincial finance annually for the construction of artificial reefs; Hebei Province has specially formulated the Measures of the Fisheries Bureau of Hebei Province for the Administration of Artificial Reefs; Guangdong and other provinces add

On March 30th, 2012, fishery administration officials were conducting fishery resources proliferation and stocking activities in the Tianjia'an segment of Huaihe River in Huainan City, Anhui Province.

the artificial reef construction into the marine environmental protection planning as an important content.

Limited fishing, banned fishing, multiplication release, and artificial reef construction are one of the main measures taken by China to promote marine and island ecological restoration, improve self-recovery capabilities, and promote a virtuous circle of ecosystem development.

In addition, China also restores small islands where the ecosystems are severely degraded, through island vegetation restoration, beach restoration, pollution treatment and utilization of renewable energy. By mangroves plantation, coral reefs cultivation, coastal wetlands reconstruction, China has rebuilt or restored the typical marine ecosystems already degraded.

Since 2010, the Ministry of Finance and the SOA have approved a total of 73 coastal waters remediation and restoration projects, and used

RMB 1.645 billion from marine space use fees assigned by the central government. In April 2013, Beidaihe comprehensive renovation and national marine security project became the first coastal waters remediation and restoration project to pass completion acceptance under the support of marine space use fees assigned by the central government.

In November 2013, China's Hebei Province promulgated and implemented the Coastline Protection and Utilization Planning of Hebei Province (2013-2020). This is China's first coastline protection and utilization planning approved by a provincial people's government. The Planning has specified the goal of coastline protection and utilization in Hebei Province: By 2020, the retention rate of natural coastlines in mainland China should be no less than 35%, and not less than 80 kilometers of damaged coastlines should be renovated and repaired. The Planning divides the coastlines into strictly protected coastlines, moderately used coastlines, and optimally used coastlines, and puts forward the objectives of upgrading, restoring and improving 134.3 kilometers of embankments, and constructing 80 kilometers of coastal landscape corridors.

Coastline protection, restoration and repair are a matter to which all coastal areas in China should attach great importance. The promulgation and implementation of the Coastline Protection and Utilization Planning of Hebei Province (2013-2020) marks that Hebei Province has stepped ahead in this regard in China. So it is expected that other coastal provinces in China will soon determine their goals in coastline development, protection and restoration so as to ensure scientific, conservative, intensive, and environmentally friendly use of coastline resources.

Speeding Up the Construction of Marine Protected Areas

In 1963, China established its first ocean-related nature reserve, the Shedao Nature Reserve, in the Bohai Sea. After that, the departments concerned have established a number of national and local marine nature reserves. A variety of typical fragile marine ecosystems, rare and endangered marine organisms, marine natural and historical sites as well as natural landscapes of significant scientific and cultural value are gradually under protection.

In 2010, China's State Council passed the China Biodiversity Conservation Strategy and Action Plan (2011-2030), and put forward China's overall objectives, strategic tasks and priority actions in biodiversity conservation in the next 20 years. In terms of marine biodiversity protection, the Action Plan has determined 3 priority areas for the protection of marine and coastal biodiversity, including the Yellow Sea protected area, East China Sea and Taiwan Strait protected area, and South China Sea protected area, and has detailed the focuses of protection in each protected area. The Action Plan has also selected the "coastal and offshore typical ecosystem protection and ecological restoration project" as one of the priority projects for biodiversity conservation, with its content including: carrying out background investigation of coastal and offshore typical ecosystems, finding out the status of all kinds of typical coastal and offshore ecosystems, and researching to develop marine ecological zoning and protection demonstration. The marine protected areas construction projects should be implemented in coastal areas with concentrated distribution of mangroves, coral reefs, seagrass beds, and coastal wetlands, as well as on islands with important ecological zones.

At present, China has basically finished establishing a marine protection network system combined by marine natural reserves and marine special reserves. By the end of 2012, China had established more than 200 various marine nature reserves of all levels (excluding Taiwan, Hongkong and Macao), including 35 national marine nature reserves. The objects of protection include rare marine animals such as spotted seals and the Chinese white dolphins, typical marine ecosystems like magroves and coal reefs, coastal geological heritages of shell dike and submarine forests and rare waterfowls such as red crowned cranes as well as their habitats.

Marine special reserve means to take effective protective measures and scientific development mode for the region with special geographical conditions, ecosystem, biological and non-biological resources and marine development and utilization of the special needs to implement special management. Compared with the great importance in protection in the nature reserve, marine special reserve focuses on development and conservation simultaneously. In 2005, the state approved the establishment of the first marine special reserve, by the end of 2012, China has established 23 international marine special reserve (not including Ocean Parks), a total area of more than 280,000 hectares.

In 2010, competent marine administrative authorities used the ocean park as a new type of marine special reserve. While the ocean park protects the special marine ecological landscape, historical and cultural relics, unique geological landscape, it gives full play to its function of ecological tourism, realizes the coordination of protection and development, and achieves win-win situation for ecological environment benefits and social and economic benefits. By the end of 2012, China has set up 18 national ocean parks.

Tianjin Ancient Coast and Wetland Nature Reserve

Important Coastal Wetlands Included in the List of Wetlands of International Importance

In 1992, China joined the Convention on Wetlands of International Importance Especially as Waterfowl Habitat. According to the Convention, China has included a number of coastal wetlands in the list of wetlands of international importance, to give full and effective protection.

Shuangtai Estuary Wetland of Liaoning Province

located in the north of the Liaodong Bay of Liaoning Province, is China's largest reed marshes in high-latitude areas with an area of about 128,000 hectares, and it belongs to an estuarine wetland. With a large area of suaeda beaches and shallow waters, the wetland serves as habitats and breeding grounds for red-crowned cranes, white cranes, larus saundersi, geese and ducks, umbrette and various passerine birds.

Dalian Spotted Seal National Nature Reserve

is located near Changxing Island of the Fuzhou Bay and 20 kilometers northwest of Dalian City of Liaoning Province, with area of 11,700 hectares. The seabed of the coast of the nature reserve is featured by steep terrains and bedrock, with water depth of 5-40 meters. The main protected species are spotted seals, which are listed as aquatic animals under the second class national protection.

Dafeng Père David's Deer National Nature Reserve (Elaphurus Davidianus)

is located in the southeast of Dafeng City of Jiangsu Province, covering an area of 78,000 hectares. The nature reserve is typical coastal wetlands, mainly including beaches, seasonal rivers, some artificial wetlands, and a large number of woodland, reed marshes, marshlands, salt barren and forest marshes.

Yancheng Nature Reserve of Jiangsu Province

is located in Yancheng City of Jiangsu Province with an area of 453,000 hectares. The reserve is located in Jianghuai Plain (plain of the Yangtze-Huai rivers) on the west coast of the Pacific Ocean. The 582-kilometer-long coastline and the vast tidal mud flat have formed China's largest tidal wetlands in coastal areas, which supporta large number of organisms, ensure the migration of millions of migratory waterfowls, and meet the needs of safe wintering of endangered species such as red-crowned cranes.

Chongming Dongtan Nature Reserve of Shanghai Municipality

is located in the low alluvial island—Chongming Dongtan in the easternmost of Chongming Island—with an area of 32,600 hectares. With the deposition of sediments from the Yangtze River, a large area of freshwater-to-brackish marshes, tidal creeks and intertidal mudflats are formed. In the reserve, there are numerous farmlands, fish ponds, crab ponds, reed fields, lush calcophic vegetation, and rich benthic fauna. It is an excellent stopover and courier station for the migration of migratory birds in springs and autumns in Asia Pacific as well as an important wintering ground for migratory birds.

Zhangjiangkou Mangrove Forest National Nature Reserve of Fujian Province

located near the estuary of Zhang River in Yunxiao County and covering a total area of 2,360 hectares, is a wetland-type nature reserve mainly protecting mangroves and wildlife. With a large area of natural mangroves distributed in the northernmost of China, the reserve is a natural mangrove community with the most species and the best mangroves in the north of the Tropic of Cancer in China.

Zhanjiang Mangrove Forest National Nature Reserve of Guangdong Province

covers an area of 20,279 hectares. The wetland is the largest coastal mangrove wetland in the southernmost of mainland China. According to preliminary investigations, there are 24 species of mangroves, 82 species of birds and abundant shallow water biological resources. When the tide ebbs, large areas of bare beaches will be exposed to provide excellent venues for waterfowls to forage and inhabit.

Shankou Mangrove Ecosystem National Nature Reserve of Guangxi Province

is located on both sides of Shatian Peninsula in Shatian Town, Hepu County, Beihai City, Guangxi Zhuang Autonomous Region, with a coastline of 50 kilometers long, a total area of 4,000 hectares, and a forest area of 806 hectares. In the reserve, there are centuries-old rhizophora stylosa and bruguiear gymnorrhiza communities, tall and concatenating, extremely rare in China; there are also many endangered wild animals such as dugongs, white dolphins, amphioxus, tachypleus tridentatus, pteria martensii, black-faced spoonbill, and larus saundersi.

Beilun Estuary National Nature Reserve of Guangxi Zhuang Autonomous Region
is located within Dongxing City and Fangcheng District of Fangchenggang City of Guangxi Zhuang Autonomous Region, with a total area of 3,000 hectares. In the reserve, there are large areas of mangroves with concatenating distribution. The mangrove plants are divided into 10 families and 13 species, forming 12 kinds of mangrove communities.

Dongzhaigang National Nature Reserve
is located in Qiongshan County of Hainan Province with an area of 5,400 hectares, mainly protecting ecosystems and wintering bird habitats in estuaries and on beaches on the northern edge of tropical zones which are dominated by mangroves. Dongzhai Harbor has 26 species of mangrove plants, and 40 species of semi-mangroves and mangrove associates, accounting for 90% of mangrove plant species in China. Dongzhai Harbor is an important stopover for many international migratory waterfowls and an important link connecting birds from different biological worlds. There are 159 species of birds inhabiting in the reserve, including 35 species listed in the Sino-Australia Agreement for the Protection of Migratory Birds (a total of 81 species of birds) and 75 species listed in the Sino-Japan Agreement on the Protection of Migratory Birds.

Constitution of a Comprehensive Monitoring System in Marine Environment

In 1984, "Nationwide Marine Pollution Monitoring Network" started to be organized in China. The hydrology and water quality of China's oceans started to be detected routinely. At present, there has been over 300 Marine Pollution Monitoring Stations established by all levels of departments that are related to the ocean. Marine Pollution Monitoring Institutions have been established in all coastal prefecture-level cities. There are also county-level marine pollution monitoring institutions established in Shandong Province, and Zhejiang Province, etc.

Means of monitoring is developing to be more stereoscopic now, which used to be ships and base stations on the shore in the early years. Buoy, airplane, satellite and radar have become normal means. The monitoring range not only covers the sea areas under the jurisdiction of our country, but also stretches to the sea areas beyond. The monitoring items include the monitoring of the trend of the marine environmental quality about the body of water, the sediment and the biological quality, the monitoring of marine environmental disasters about the seawater encroachment and soil salinization, the emergency monitoring of emergencies such as red tide and oil spilling, the health monitoring of marine ecology system, and the monitoring specialized on some special areas such as drain outlets to the

Sun Jie from the Chinese Academy of Surveying and Mapping, and the self-developed, with international advanced level "Low-altitude UAV Remote Monitoring System"

sea, estuaries, dump areas, vacation areas, engineering construction areas, bathing beaches, mariculture areas and marine reserves.

In 2004, China launched establishment of nationwide inshore marine ecosystem monitoring areas. 18 ecosystem monitoring areas were established in typical marine ecosystems and ecological sensitive areas including estuaries, coastal wetlands, mangroves, coral reefs, seagrass beds and bays, to monitor environmental indicators, biological indicators and ecological pressure indicators, evaluate the health and safety conditions in the marine ecosystems, screen the major marine ecological problems and causes, and provide support for the integrated management of the environment in coastal zones.

In recent years, China has also organized special monitoring, such as "monitoring of the status and trends of marine environmental quality," "monitoring in offshore red tide monitoring zones," "monitoring in offshore ecological monitoring zones," "monitoring in key sewage outfalls and adjacent waters," "offshore environmental quality monitoring," and "environmental monitoring in fishing waters", and has regularly issued the State of Environment in China, the China Offshore Environmental Quality Bulletin, the China Fishery Ecological Environment Bulletin, and the bulletin of special monitoring of the marine environment to provide a basis for mastering the current status and trends of the marine environment in China.

To Achieve Normalization of Marine Environmental Protection Law Enforcement

"Blue" series series of special law enforcement actions implemented fromfrom 2009, the action positioned for thespeciallaw enforcement operations ofof marine environmental protection, preventing the marine engineering construction construction projects of pollution damage to the marine environment is thekey, strengthening supervision and inspection, and crack down on and dealt with serious marine environmental violations according to law for offshore oil exploration and exploitation, ocean dumping, marine nature reserve, special marine reserves, marine ecological monitoring area, the main pollutant emission export and other fields,

According to the unified deployment,, law enforcement on marine marine engineering environmental protection law enforcement will focus on completing the file registration of the marine engineering project, implementing implementing the dynamic environmental protection law enforcement through new construction, reconstruction and expansion and investigating and punishing illegal acts in all kinds of marine projects. The main tasks of marine petroleum exploration and environmental protection law enforcement are supervising and examiningthe activities of platforms and ships that are engaged in marine petroleum exploration, at the same time establishing regular cruise system in offshore petroleum exploration with focus on Bo Sea, monitoring marine petroleum platforms and nearby areas and implementing dynamic supervision of platforms'discharge of pollutant. Marine dumping law enforcement emphasizes strengthening the cruise supervision of marine dumping areas and temporary dumping areas, establishing and implementing verification system of loaded pollutant that has been permitted to dump and supervising and examining marine platforms' activities of dumping in remote areas. Marine ecological

On November 3rd, 2010, fishery administration department in Qionghai City, Hainan Province, was carrying out special law enforcement operation under the theme "Cherish turtles, Protect the Sea" in Tanmen Port. The pictured showed the law enforcement staffs were releasing the confiscated turtles.

protection law enforcement will focus on the inspection of national and provincial marine nature reserves, national marine special reserves andmonitoring area of marine ecology. For the key dumping areas, actions will be taken to strengthen the supervision of sewage outfalls near lands and deep seas and build up environmental supervision, law enforcement and management information report system in main sewage outfalls.

In 2013, marine surveillance institutions at all levels organized byChina Marine Surveillance to implement Blue Ocean 2013special enforcement action achieved great success. By the end of 2013, there were altogether 152 registered cases, 138 issued administrative punishment decision bookand 137 settledcases, and confiscated 11,238,000 yuan.

According to the State Oceanic Administration, in the law enforcement of ocean engineering construction project, the whole process of supervision system in marine engineering construction project has been established gradually. Marine surveillance institutions at all levels strictly implement the responsibility of day-to-day supervision of marine engineering project,comprehensively track certain marine areas' construction projects, and constantly improve the registration filing work. Annual inspection project basically has been filled and registered to basically realizethe whole process supervision from construction to operation of marine engineering construction project.

In the ocean dumping law enforcement, surveillance institutions at all levels always keep themselves under high pressure with full attention on major areas and dredging project. They rely on ocean dumping recorders, satellites and video monitors and other high-tech means to carry out specific law enforcement through a variety of ways such as ocean cruise, air patrol, nocturnal ambush, seashore watching and board examining. The number of annual total dumping cases was 98, a lower incidence than 2012 apparently.

According to the relevant responsible person in China Marine Surveillance,China Marine Surveillance will continue to rely on platforms of Blue Ocean series to further expand the field of law enforcement and comprehensively promote the marine environmental protection law enforcement work.

In the first half of 2013, the China Marine Surveillance organized 3 Marine Corps to carry out overall law enforcement inspection on marine oil exploration. From March 15thto June 5th 2013, North Ocean Corps launched the eighteenth law enforcement inspection of regular cruise on Bo Sea. The cruise covered 1841 nautical miles, 2500km from the coastline and in 177 hours. In the meantime, law enforcement officers patrolled 36 marine oil and mining areas, 103 platforms, 6 oil preserving ships, ecological monitoring areas and 7 natural reserves, 7 dumping areas, 2 sea sand mining areas and took 130 photos.

Bo Sea tour group carried out a cruise inspection on targets like offshore platforms, storage tankers and oil pipelines after the 19-3 oilfield's reproduction in Penglai.

At the same time, the South Marine Corps and East Marine Corps also conducted regular cruise law enforcement inspectionon their separate oil and gas exploration areas.

During the cruise in law enforcement inspection, the Marine Surveillance found 4 suspected illegal cases. Some offshore platforms constructed platforms when the marine environmental impact report hadn't been approved by the Oceanic Administrative Department; some platform's environmental protection facilities were put into use without approval of Oceanic Administrative Department.

In the cruise law enforcement inspection, some of the offshore oil platforms hadno logo; sewage, drilling fluid and drilling cuttings hadn't been submitted for inspection in time; people didn't hand in the quarterly pollution preventing report of offshore oil fields; people drilled platforms without oil spill contingency plan; life sewage treatment devices were overloaded with excessive output; labels on the environmental protection facilities were not clear, oil spill emergency equipment was not preserved according to standards and people in the terminal treatment plant did not fill in the pollution preventing records.

The China Marine SurveillanceDepartment have dealt with those issues seriously according to law.

In the sea areas under the jurisdiction of Shandong Province, there are altogether 589 islands, of which are 32 residential islands and 557 uninhabited islands, mostly in the coastal areas. With the rapid development of marine economy, the contradiction between protection and development of island resources is becoming increasingly prominent. Part of the island's ecological environment has been destroyed.The developing and utilizing form of the island is still relatively rough.

According to this, Shandong province has implemented a strict "grid" management, a regular inspection system called "one period for one island". In each quarter, Marine Surveillance institutions in cities and towns should conduct cruise law enforcement on islands with residents and islands without residents but development and utilization activities, strengthen theprotection of island resources and standardize the development and utilization activities. For islands with residents, they should focus on the examination of unauthorized reclamation, illegal sewage and unauthorized excavation of sand. As for those islands with no residents, they should focus on the examination of illegal reclamation, illegal construction, exploitation and utilization of island that violate specific implementation program. For territorialsea points, they should focus on checking whether there are unauthorized engineering projects and other activities that may change the regional topography and landforms and whether the symbol of the territorial sea point has been damaged or moved or not.

In 2013, cities and countries in Shandong province have taken regular law enforcement, special law enforcement and joint law enforcement 472 times one by one, dispatched law enforcement boats for 93 cruises, taken more than 6500 photos and sailed 5000 nautical miles.

Best Practice of Marine Nature Reserve

Shankou National Mangrove Nature Reserve of Guangxi Taking the Path of Sustainable Development

Shankou mangrove nature reserve of Guangxi in China located in the jurisdiction of Shankou town, Hepucounty in Guangxi Zhuang Autonomous Region, it covers an area of 8000 hectares, including an area of 806.2 hectares forests. In September 1990 by the State Council approved the establishment of Shankou mangrove nature reserve, one of the five first national marine reserves. In March 1992, reserve management station was established in of Ma Anpeninsula of the rowan. In 1993, it joint in Chinese biosphere reserve network. In 1994, it was listed as important wet land in China. In May 1997, it established sister reserve relationship with National Estuarine Research Reserve in Luke bay in Florida, the United States. In January 2000, it joint in World Biosphere of UNESCO. In February 2, 2002, it was listed as International Important Wetland Protectorate.

Mangrove forests in Shankou Mangrove Reserve is the typical representative of mangrove forests in Chinese mainland coastline, including ten varieties of true mangrove, such as Bruguieragymnorrhiza, Kandeliacandel and so on, and five varieties of semi-mangrove, such as

Acrostichumaureum, Section park and so on.

The mangrove forests with luxuriant foliage provide an ideal habitat for marine life and birds.The reserve has 96 kinds of phytoplankton, benthic diatoms of 158 kinds of fish, 82 kinds, 90 kinds of shrimp and crab shell, 61 species, 132 kinds of birds, 258 species of insects, 26 species of other animals.The sea nearby is the habit of national first class protected animal "Mermaid" dugong, the Chinese white dolphin, amphioxus and Tachypleustridentatus and other rare marine animals, and is also the breeding area of the mother of pearl in Hepu.

Reserve adheres to the protection policyof "conservation based, moderate development, of sustainable development", and keeps close cooperation withdomestic Surgery Institute, colleges and universities, carries out scientific research development of mangroves, explores the methods of comprehensive development and sustainable utilization of Mangrove Resources in a reasonable way, attempts to establish the reserve as the base of mangrove resources conservation, research,teaching, international communication, development and tourism.

After the establishment of reserve, the government of the Guangxi autonomous region based on "nature reserve regulations of people's Republic of China" and "marine nature reserve management measures", combined with the characteristics of Shankou mangrove reserve,formulated and promulgated the "Shankou mangrove ecological nature reserve management approach", Hepu County People's Governmentwhere the reserve located in have specialized issued "announcement on the strengthening of national Shankou mangrove ecological nature reserve", these regulations provides a legal basis for the management of reserve.

Expect the establishment of a marine monitoring team to carry out law enforcement management work, the reserve has set up theYingluo and Yongan supervision and management station, and hired rural village cadres as a part-time management personnel, established a management system from the reserve management to the village committee.

Winding wood bridge in Guangxi's Shankou Mangrove Reserve

After the establishment of the reserve, the residents' former mode of production and life has been affected to a certain extent. How to make people aware of the significance of the establishment of the reserve andconsciously support the work of the reserve is the emphasis and difficulty of it. The management office of the Shankou Reserve has penetrated deeply into towns and coastal villages to promote laws and rules, expand social understanding of the reserve, improve people's awareness of protection and make people consciously participate in the protection of the mangrove ecosystem using documents, wall papers, slogans, banners, advertisements and seminars attended by village officials.

On 15th September, 1996, a strong typhoon attacked the Yingluo Bay. Over 50 fishing boats outside the forests were overturned instantly and 22 lives were claimed while more than 30 boats and the crew in the tidal creek inside the forests remained safe and sound. This event has greatly educated people and the mangrove protection has become a conscious action of the masses. For example, Tan, head of the Village Committee, Baisha, has mobilized the masses to plant mangroves for more than 30 hectares, for the purpose of protecting the village's shrimp farm.

The reserve has been integrated with its community as well. The construction of the reserve and the community's economy has been developing together. The constant development of the reserve has brought benefits to the surrounding villages and the masses. The construction of the reserve provides infrastructure such as electricity and traffic lights for the nearby communities. The blossom of ecological tourism industry has driven the development of individual passenger transportation and other related industries. Meanwhile, the breeding industry has been developed thanks to the natural shelter provided by the mangroves. At the bank of the mangroves near Dangjiang and Xichang, Hepu, people who graze chicken, duck and grabs can be seen at times. Mr. Chen, who is a farmer, specifically did experiments that ducks in the mangroves lay 7 to 8 times of eggs than that on the shore. Nowadays, the mangrove has become more than the name of a kind of plant but also of a place, a company or a shop. There have already been fruit, a pearl field, a restaurant and a middle school with a name ofMangrove. Shankou has been well-known for mangroves in Guangxi District.

Shankou Reserve develops ecological tourism using its wonderful, peaceful and beautiful natural environment, which provides visitors with knowledge about the ecosystem, coastal geomorphology, marine biology etc. in a direct and emotional way. It also makes tourists get educated about marine ecological protection and science in the playing.The reserve is named as "Beihai Science Education Base" and "Guangxi Science Education Base". With the improvement of tourism conditions, the number of visitors has increased year by year.

In the past, villagers had weak consciousness towards environment protection and some villagers once cut down mangroves. After the 20-year promotion and education, protecting the mangroves has become a common will between management departments and the community people. Shankou

Mangroves Reserve has embarked on a win-win and healthy development road of community development and mangroves protection. This has also become a microcosm of the development of marine protected areas in China.

Harmonious Relationship between Nature and Human Beings : Zhuhai Qi'ao-Dangan Island Nature Reserve

Zhuhai Qi'ao-Dangan Island Nature Reserve, established with the approval by Guangdong Provincial People's Government in November 2004, is composed by the combination of Zhuhai Dangan Island Macaque Provincial Nature Reserve and Zhuhai Qi'ao Island Mangrove Municipal Nature Reserve.

The Mangrove Nature Reserve covers a total area of 5,103.77 hectares, with 533.3 hectares of mangroves, which boasts 695 species of vascular plants and 347 species of wild animals, including 15 species of true mangrove plants and 9 species of semi-mangrove plants.

The mangroves growing on the beach not only serve as a marine forest to break wind, fix sands, prevent waves and protect dikes, but also act as a good place for birds and marine organisms to inhabit and multiply.

As one of three major migratory paths in China, this nature reserve serves as a habitat for more than 90 species of migratory birds in autumns and winters.

The Macaque Nature Reserve is located at the junction of the Lingdingyang Bay in the southern Pearl River Delta and the South China Sea, with a total area of 2,270 hectares. The island is rich in flora and fauna, with 438 species of vascular plants, 85 species of wild animals, including 12 species of rare plants and animals; it is also home to 3 species of plants enjoying third-class national protection, 1 species of animals enjoying first-class national protection, and 9 species of animals enjoying second-class national protection. Due to the sea wind, the bonsai plants on the island are

On July 30th, 2012, Liu Qingwei, the leader of the Dangan Island Macaque Reserve fed macaques every day. He has worked here for 23 years, and macaques have become his friends.

in various natural shapes.

The number of macaques on the island has increased from less than 300 in 1982 to over 1,300 now, of which nearly 100 are artificially domesticated, and often play with people, which is very interesting.

By 2014, Liu Qingwei has been fraternizing with the macaques on Dangan Island for 25 years. For decades, this ranger and caregiver at the Dangan Macaque Protection Station of Zhuhai Dangan Island Provincial Nature Reserve has been together with macaques for a far more time than with humans.

In 1989, Liu Qingwei completed his term of service in Zhuhai garrison. Military leaders asked him about going to work on Dangan Island, where a protected area was newly established, in need of forest rangers, but life would be very hard. Liu Qingwei replied: "I'll go wherever the army arranged, and I'm not afraid of hard work."

"I really did not expect that life on the island would be so hard," said Liu Qingwei. "It took seven to eight hours to arrive at the island by boat. When the boat arrived on the shore, I only saw one person on the pier. Miscanthus and trees are everywhere. Only dozens of fishermen households temporarily lived there by fishing."

After landing on the island, Liu Qingwei returns to Zhuhai once every two or three years. His routine job is to feed the macaques in the protected area and record the growth process and eating habits of them; and prevent lawbreakers from hunting macaques and stealing valuable trees such as podocarpus macrophyllus.

To protect the forests on the island is one of Liu Qingwei's responsibilities. He said, "I'll do it all my life. I won't change my mind."

However, the job that he has promised to undertake all his life is not only a hard and life-threatening thing, but also a pain that can never heal.

During his first investigation around the island in 2013, Liu Qingwei encountered the 8th typhoon that year, where waves might overturn his boat at any time. Liu Qingwei later said: "I got wet all over by the seawater. The waves seemed to send my boat to heaven. At the beginning, I was very nervous. Then I calmed myself down and finally got through it and sailed back to the island."

The disease of his son, named Cong Cong, is Liu Qingwei's pain that can never heal. Since he landed on the Dangan Island in 1989, Liu Qingwei could rarely return to Zhuhai. Before the Spring Festival in 1991, since he missed his 4-month-old son, Cong Cong, very much, he asked his wife to take their kid to the island to celebrate the Spring Festival there. After only a few days' reunion, his son suddenly had a high fever. Because of erratic weather and storms, there was no boat available for use. So after over 10 days, they finally managed to get on a boat heading to Zhuhai. They went to a large hospital in Zhuhai for an examination. Upon examination, the doctor scolded the couple on the spot: "The child had such a high fever. Why didn't you send him here earlier?" The child was diagnosed with cerebral palsy.

For Liu Qingwei, macaques are an interconnected part of his life. He said, "They are just like my brothers." Due to long-term residence on the island, Liu Qingwei suffered from gout. There was once a period when he was transferred back to Zhuhai. But a few days later, he missed the macaques on the island and was really worried: "Did they eat vegetables planted by fishermen? Did they fight against each other?"

Liu Qingwei has conducted long-term observation and record of the habits of macaques and accumulated a large amount of research data and information. He said: "As I watch them and see their emotions every day, I have feelings forthem and can understand their language."

It was due to such "feelings" that the macaques saved Liu Qingwei's life. One day, a monkey came to him and caught his fish on the back. They money was hurt by fish and scratched in the dirt. Then it grabbed a plant of dripping Guanyin on the roadside and wiped paws with juice. After a while, the wound was healed and the money stopped scratching.

One day after that, Liu Qingwei accidentally broke a hornet's nest, and swarms of hornets chased him ad stung him. He was so painful that he suddenly thought of the dripping Guanyin once used by the macaques. He tried it and it indeed worked. 90% of those who were stung by hornets would be dead. It was the macaques and dripping Guanyin that saved his life.

"Podocarpus macrophyllus and Chinese littleleaf box in the protected area are very precious plants. While protecting them on the island, I almost lost my life." Liu Qingwei said, "There are countless thrilling scenes of fighting with thieves."

One day, when Liu Qingwei took the fishermen's boat to perform inspection on the Erzhou Island near the Dangan Island, he encountered thieves. He immediately reported to his superiors by telephone, seeking for additional support. The thieves saw him and decided to hit his boat. When it was about to hit, the thieves' speedboat was raised up by waves just above his head and the propeller passed over him heavily. In case that the speedboat was raised a few centimeters lower, it would kill him.

In June 2012, his wife and son moved to the Dangan Island to live with Liu Qingwei. In the past, the family could only "meet once every three years". When they made phone calls, there may be no signal at any time. Now they can finally live together.

Now, the number of macaques has increased from over 200 to more than 1,000.

From the story of Liu Qingwei, we may see that China has made long-term and arduous efforts to protect the natural resources in oceans and on islands.

CHINA'S MARINE CONSERVATION AND DEVELOPMENT

FUTURE OF MARINE DEVELOPMENT
——CHINA'S SUSTAINABLE MARINE DEVELOPMENT STRATEGIES

Progress of Sustainable Marine Development

Sustainable Marine Development in the World

The early writings about environmental movement and environmental protection have presented us the environmental hazards and enabled us to recognize the existence of environmental problems. They have formed the atmosphere in which modern society discusses and solves environmental problems, and laid a foundation for the proposal of the concept of world sustainable development.

Environmental pollution entered public life as a social issue in 1962 when the book Silent Spring was published. Through revealing the hazards of pesticides, the book sparked a heated debate on environmental issues. Ten years later, the Club of Rome published The Limits to Growth, predicting that the limit of human economic growth would arrive in the near future, so that people realized for the first time the limitations of the supporting role of the Earth system, challenging the traditional concept of economic growth. These two enlightening works have triggered a series of environmental protection works and activities, gradually forming the public opinions of pollution control and environmental protection.

On this basis, in 1987, the World Commission on Environment and Development put forward the concept of sustainable development in the Our Common Future, holding that the development of contemporary

United Nations Conference on Environment and Development, 1992, Rio de Janeiro

society should "meet the needs of the present without compromising the ability of future generations to meet their needs." Later, the concept of sustainable development was further confirmed and developed on the three World Summits on Sustainable Development, and was put into action plan through outcome documents of the summits, including Agenda 21, Plan of Implementation of the World Summit on Sustainable Development, and The Future We Want. Oceans are an important part of the world's sustainable development.

1992 United Nations Conference on Environment and Development in Rio

Sustainable development was established as the guiding principle of the world's economic, social and environmental development on the United Nations Conference on Environment and Development (UNCED), also known as the "Earth Summit", held in Rio in 1992. The Conference aimed to review the global environmental protection process in the past 20 years after the convening of the first UNCED, and urge governments and the public to take positive measures to coordinate with each other to prevent environmental pollution and ecological degradation and make joint efforts to protect the human environment.

The Conference passed important documents, such as Rio Declaration on Environment and Development, Agenda 21, and The Declaration of Principles on Forests, and opened for signature of the United Nations Framework Convention on Climate Change and the United Nations Convention on Biological Diversity, reflecting a global consensus and political commitment at the highest level in the field of environment and development cooperation. The important document of Rio Declaration on Environment and Development adopted by the Conference has clearly put forward the implementation of sustainable development strategies, provided several important principles of common but differentiated responsibilities and priorities to consider the situations and needs of developing countries, and put forward the main principles and systems of environmental management. The Rio Declaration on Environment and Development signals that the concept of sustainable development has been recognized as an international consensus, and an order for international cooperation in the field of environment and development has been established.

Another important outcome of the Rio summit was the adoption of Agenda 21, which put forward a blueprint for action to achieve global sustainable development in the 21st century. It serves as a global framework for us to take measures to safeguard our common future and has an epoch-making significance.

Chapter 17 of the Agenda 21 specifically explains the initiatives to "protect oceans and seas including enclosed and semi-enclosed seas and coastal areas, and protect, rationally use and develop the biological resources," and explains the policies and measures to promote the sustainable development of oceans. It has an important guiding significance for the marine environment protection and resources management in various countries.

2002 World Summit on Sustainable Development in Johannesburg, South Africa

The World Summit on Sustainable Development took place in Johannesburg, South Africa, from August 26th to September 4th of 2002. The Summit adopted two resolutions, the Johannesburg Declaration on Sustainable Development and the Plan of Implementation of the World Summit on Sustainable Development.

It stipulates that oceans, seas, islands and coastal areas form an integrated and essential component of the Earth's ecosystem and are critical for global food security and for sustaining economic prosperity and the well-being of many national economies, particularly in developing countries. Ensuring the sustainable development of the oceans requires effective coordination and cooperation, including at the global and regional levels, between relevant bodies, and actions at all levels.

2012 United Nations Conference on Sustainable Development (Rio +20)

The United Nations Conference on Sustainable Development took place in Brazil in June 2012. Since it was held about 20 years after the Rio Summit in 1992 and intended to reaffirm the principles of sustainable development established by the Rio Summit, the Conference was also known as "Rio +20."

The Rio +20 Conference adopted the outcome document entitled The Future We Want. The document reaffirmed the important principles on sustainable development determined at the Rio Summit and continued to take the Rio Principles as a foundation for cooperation, coordination and implementation of the agreed commitments between international communities. Meanwhile, the Rio +20 summit also assessed the achievements and deficiencies faced in the process of achieving sustainable development, summed up the new challenges faced by countries in the sustainable development process, and proposed measures to deal with all kinds of challenges.

Oceans are one of the seven "Rio +20" key areas (employment, energy, cities, food, water, oceans, disasters) defined by the United Nations. The Rio +20 summit has pointed out the crucial role of oceans in the world's sustainable development and regarded oceans as an important area in the framework for action and follow-up actions in the outcome document of "Rio +20". The outcome document The Future We Want holds that "oceans, seas and coastal areas form an integrated and essential component of the Earth's ecosystem and are critical to sustaining it."

With the promotion of three World Conferences on Sustainable Development, through the global practice of the principles of sustainable development and the implementation of the Agenda 21, the consensus for sustainable development has gained considerable development in various fields and the connotation of it has been continuously enriched. In the field of sustainable marine development, as countries intensify their efforts in marine development and enhance their understanding of the oceans through marine scientific research, topics such as blue economy, high sea protection, and climate change are becoming new hot spots.

Sustainable Marine Development in China

In March 2014, four marine nonprofit industry research and special funding projects undertaken by China's Tianjin Municipality were launched. The four projects involved multiple fields, including desalination, marine chemicals, marine monitoring, and marine environmental protection.

To promote the construction of marine economic and scientific development demonstration areas in Tianjin, SOA has listed Tianjin as the construction unit of the annual marine nonprofit special funding projects and spent a total of nearly RMB 80 million funds to support the projects.

Environmental protection and sustainable development are important policies of the Chinese government. Since the 1970s, China has taken the "reduction of environmental pollution and protection of natural resources" as a priority area of national policies. Especially in the 1990s, environmental protection was listed as a basic national policy.

The Chinese government has actively involved in the process of sustainable development in the world. It has attended the 1972 Stockholm Conference on the Human Environment, the 1992 Rio Conference on Environment and Development, the 2002 Johannesburg World Summit on Sustainable Development and the 2012 Rio +20 Conference. In 1994, the Chinese government issued the China's Agenda 21. In 1996, sustainable development was formally identified as one of China's basic development strategies, marking the transition of sustainable development from a scientific consensus to an important part and concrete action of the government work.

Since the implementation of the China Ocean Agenda 21 in 1996, China's sustainable marine development has gone through nearly 20 years, which was the transition period of China's economic and social development. During that period, philosophies, development concepts and strategic visions that embody the thinking of sustainable development, such as well-off society, harmonious society, environmentally-friendly and resource-saving society, and ecological civilization, were proposed to speed up the process of sustainable development in China. In the 21st century, the Chinese government pays more attention to the development of the marine industry, which gradually becomes a priority area of the national economic and social development. With the continuous improvement of sustainable development policies, the sustainability of the oceans is steadily improved.

The China's Agenda 21 declares that China takes the road of sustainable development; the China Ocean Agenda 21 establishes sustainable marine development as the guideline for the development of China seas; the White Paper on the Development of China's Marine Programs explains the basic idea of the sustainable development strategies of China seas, accordingly puts forward the basic strategies and principles of China's marine development, and establishes the basic principles of

On June 10th, 2010, the "Training to Address Climate Change" was held in Beijing jointly hosted by the China Agenda 21 Management Center, American Climate Project Organization and Sino-US Sustainable Development Center.

sustainable development strategies in China's marine industry development. On this basis, the 10th, 11th and 12th five-year guidelines have made important deployment of China's marine industry and implemented the sustainable marine development strategies in the government's work.

China's Agenda 21

Following the United Nations Conference on Environment and Development in 1992, the Chinese government issued China's Agenda 21 — White Paper on China's Population, Environment and Development in the 21st Century. Prepared in accordance with China's national conditions, the China's Agenda 21 is a policy document echoing the United Nations Agenda 21 and is an important basis for China to implement the principles of sustainable development. Meanwhile, it has extensively included the action plans, with strong operability, being and to be implemented by various government departments, and is also a core file for China to construct a sustainable development path.

In the preface, the White Paper proposes at the outset a strategic position of sustainable development: It is China's demand and inevitable choice to develop and implement the China Agenda 21 and to take the road of sustainable development in the future and the next century. In terms of content, the China's Agenda 21 consists of 20 chapters and 78 program areas, involving four aspects of overall strategies of sustainable development, sustainable development of society, sustainable economic development, and rational use and protection of resources and the environment, and covering Chinese population, economy, society, resources, environment, etc. In terms of rational use and protection of resources and the environment, the White Paper takes "sustainable development and protection of marine resources" as an important area of action, and proposes to improve mechanisms for integrated management of national marine resources, protect marine biological resources, develop and protect coastal and island resources, and construct marine science and technology demonstration projects.

As China's first policy document to announce the sustainable development strategies, the China Agenda 21 contains the content of marine resources development and protection and takes the marine industry as an important area of action for sustainable development in China. It is a high-level policy basis for China to implement sustainable marine development.

China Ocean Agenda 21

In order to better implement the China Agenda 21 in the marine sector and promote sustainable development and utilization of the oceans, the Chinese government issued the China Ocean Agenda 21 in 1996. The agenda, an in-depth and concrete manifestation of the China Agenda 21 in the marine sector, has clarified the basic strategies, strategic objectives, basic measures, and main areas of action for sustainable marine development, and can be regarded as policy guidelines for sustainable marine development and exploitation.

The China Ocean Agenda 21 takes sustainable marine utilization and coordinated development of the marine industry as the guiding principle of China's marine work in the 21st century. In order to implement this guiding principle, the document continues to raise a number of basic measures: guide establishment and development of the marine industry according to the principles of sustainable development; link the marine development with social and economic sustainable development in coastal areas, and gradually remove the major constraints to social and economic development through planned and targeted marine development activities; promote sustainable development of coastal islands, and take into account development and utilization of coastal islands, economic construction of the entire nation, as well as sustainable development of coastal areas; protect sustainable use of marine living resources; promote sustainable marine development and utilization through scientific and technological progress; establish comprehensive marine management systems; protect the marine environment; strengthen marine observation, forecasting, early warning and disaster reduction; enhance international cooperation; promote public participation in the marine industry.

The China Ocean Agenda 21 consists of actions such as conservation and management of living marine resources, construction of marine nature reserves, and control of land-sourced pollution. It acts as an integrated program in the field of rational use and protection of marine resources and environment.

White Paper on the Development of China's Marine Programs

On November 19th, 2013, the "China Marine Surveillance 1093" law enforcement speedboat built and coded by China Marine Surveillance and freely provided to marine surveillance corps at basic levels, was formally delivered to China Marine Surveillance Pulandian Corps in northeast China.

The "China Marine Surveillance 1093" law enforcement speedboat is 18.77 meters long, 4.78 meters wide, and 2.70 meters deep, with a draft of 1.082 meters. With at most 15 people on board and at a speed of 45 sea miles, it has endurance of 300 sea miles. This speedboatis equipped with advanced power equipment, modern communication and navigation equipment and multi-functional investigation and evidence collection equipment. It is characterized by strong mobility, fast speed and long endurance, and mainly used for island protection and management, use and management of the waters, cruise and marine environment protection.

The delivery of this speedboatconfirms from one side that China is accelerating itspace of marine protection.

As early as May 28th, 1998, China's State Council Information Office published the White Paper on the Development of China's Marine Programs, to further promote implementation of sustainable marine development strategies.

This 12,000-word White Paper elaborates China's marine programs from six aspects: sustainable marine development strategy, rational development and utilization of marine resources, protection and preservation of the marine environment, development of oceanographic science, technology and education, implementation of comprehensive marine management, and international cooperation in maritime affairs.

The White Paper points out that in order to develop national economy, China, as a major developing country with a long coastline, must take marine development and protection as a long-term strategic task and conduct comprehensive development and utilization according to the bearing capacity of marine resources so as to promote the coordinated development of the marine industry.

The White Paper has explained the basic idea of sustainable marine development strategies and proposed the basic policies and principles for development of China's marine programs: safeguarding the new international marine order; overall planning for marine development and control; rationally utilizing marine resources and promoting coordinated development of the marine industries; synchronizing planning and implementation of development of marine resources and protection of the marine environment; reinforcing oceanographic technology research and development; setting up a comprehensive marine management system; actively participating in international cooperation in the field of marine development.

The White Paper has affirmed China's rational development and utilization of marine resources and its role in promoting coordinated development of industries; China has continued to transform traditional industries such as maritime fishing, transportation, sea salt industry; vigorously develop emerging industries such as mariculture industry, oil and gas industry, and pharmaceutical industry; actively explore new marine resources that can be developed, and promote the formation and development of potential marine industries such as deep-sea mining, comprehensive utilization of seawater, power generation with ocean energy.

The White Paper has also introduced the efforts China has made to protect and preserve marine environment and described that China has gradually established marine environmental protection agencies and marine environmental protection regulations and that China has made great progress in control of land-sourced pollution, prevention of ship and port pollution, prevention of petroleum development pollution, and strengthened management of ocean dumping.

On the whole, the White Paper has reaffirmed the sustainable marine development strategies and introduced the specific actions for marine resources and environment protection in China. Thus, it is an important file with a comprehensive clarification of sustainable marine development strategies and actions.

Plans and Programs for National Economic and Social Development

The Outline of the Tenth Five-Year Plan for National Economic and Social Development (2001) has pointed out the direction for development of marine industries: "Increase survey, development, protection and management of marine resources, strengthen research and development of marine using technologies, and develop marine industries."

The Outline of the Eleventh Five-Year Plan for National Economic and Social Development (2006) has dealt with oceans in a separate chapter for the first time, specifying the need to strengthen maritime awareness, safeguard maritime rights and interests, protect marine ecology, develop marine resources, implement comprehensive marine management, and promote development of marine economy. In addition, it has proposed specific measures to protect marine environment, such as governing major marine environment of the Bohai Sea, the Yangtze River Estuary, and the Pearl River Estuary, and protecting coastal ecosystems of mangroves, coastal wetlands and coral reefs, and has significantly strengthened the policy guidance of marine environment protection.

Chapter 14 of the Outline of the Twelfth Five-Year Plan for National Economic and Social Development (2011) deals with "promoting the development of maritime economy", which is further elaborated in two sections of "optimizing marine industrial structure" and "strengthening comprehensive marine management".

> **Twelfth Five Year Plan for National Economic and Social Development about oceans**
>
> Stick to coordinated development of land and sea, formulate and implement marine development strategies, and improve capabilities for marine development, control and integrated management. Scientifically plan development of marine economy, rationally exploit and use marine resources, actively develop industries of offshore oil and gas, marine transportation, marine fisheries, and coastal tourism, and foster emerging industries of marine biomedicine, comprehensive utilization of seawater, and marine engineering equipment manufacturing.
>
> Strengthen research and development of basic, forward-looking, and key marine technologies, improve the marine science and technology level, and enhance the ability of marine

development and utilization. Deepen the integration of resources at ports and along shorelines and optimize the port layout. Develop and implement the planning of main functional areas in oceans, and optimize marine economic space layout. Promote development of marine economy in Shandong, Zhejiang and Guangdong.

Strengthen coordination and improve marine management systems. Strengthen ocean and island management, improve market mechanism of sea use rights, promote island protection and utilization, and support development of outlying islands. Coordinate protection of marine environment and prevention of land-sourced pollution, and strengthen protection and restoration of marine ecosystems. Control overdevelopment of offshore resources, strengthen reclamation management, and strictly regulate use of uninhabited islands. Improve marine disaster prevention and mitigation systems, and enhance capabilities to respond to maritime emergencies.

Strengthen comprehensive marine survey and mapping, and actively carry out scientific investigations of polar regions and oceans. Improve sea-related laws, regulations and policies, increase marine law enforcement, and maintain marine resources development order. Strengthen bilateral and multilateral negotiations about ocean affairs, actively participate in international maritime affairs, guarantee the safety of maritime transport channels, and safeguard China's maritime interests.

Organizer and Participant of the World's Marine Conservation Actions

China supports and actively participates in international marine conservation affairs, and has joined and approved a number of international conventions on marine environmental protection and fulfilled relevant international obligations. For example, the "Convention on Wetlands of International Importance Especially as Waterfowl Habitat" reuiqres the state parties to designate suitable wetlands within their territories into the "List of

Wetlands of International Importance" and carry out sufficient and effective protection. China has included the Hainan Dongzhaigang Mangrove Nature Rserve, Shanghai Chongming Dongtan Nature Reserve, Liaoning Spotted Seal National Nature Reserve, Zhanjiang Mangrove Forest National Nature Reserve, National Huidong Sea Turtle Reserve, Shankou Mangrove National Nature Rserve, Yancheng Coastal Tidal Wetland, Shuangtai Estuary Wetland of Liaoning Province, Beilun Estuary National Nature Reserve, and Zhangjiangkou Mangrove Forest National Nature Reserve into the "List of Wetlands of International Importance" and are managed with corresponding measures.

	International Conventions Related to Marine Ecosystem Protection that China has Participated In
1.	United Nations Convention on the Law of the Sea
2.	Convention on Biological Diversity
3.	Convention on the Prevention of Marine Pollution by Dumping of Wastes and Other Matter
4.	Convention on Wetlands of International Importance Especially as Waterfowl Habitat
5.	Protocol on Environmental Protection to the Antarctic Treaty
6.	International Convention on Civil Liability for Oil Pollution Damage of 1969
7.	International Convention for the Prevention of Pollution from Ships of 1973/1978
8.	Basel Convention on the Control of Transboundary Movements of Hazardous Wastes and Their Disposal
9.	International Convention on Oil Pollution Preparedness, Response and Cooperation of 1990
10.	Protocol of 1992 to Amend the International Convention on Civil Liability for Oil Pollution Damage of 1969

China has participated in international cooperation in marine ecosystem protection in three major ways: first, bilateral or multilateral intergovernmental cooperation with neighboring countries or other coastal countries; second, cooperation with international organizations such as UNEP, the United Nations Development Programme (UNDP), and IMO; third, various forms of cooperation with marine environment research institutes and universities in coastal countries.

The international marine ecosystem protection projects that China has participated in mainly include the "Project to Reverse the Trend of Environmental Degradation in the South China Sea and the Gulf of Thailand", the "Action Plan for Environmental Protection, Management and Development in the Northwest Pacific Ocean and Its Coastal Areas", the "Yellow Sea Large Marine Ecosystem Project", and the "South China Sea Biodiversity Management Project".

The Project to Reverse the Trend of Environmental Degradation in the South China Sea and the Gulf of Thailand

It is a large-scale regional cooperation project for marine environmental protection, co-sponsored by China, Vietnam, Cambodia, Thailand, Malaysia, Indonesia, and Philippines around the South China Sea, organized and implemented by UNEP, and funded by the Global Environment Facility. The overall objective of the project is: to create the atmosphere of cooperation and participation at the regional level, and solve the environmental problems of the South China Sea; to foster and encourage the parties to cooperate and participate at all levels; to strengthen the participating countries' capabilities to integrate the environmental considerations into their national development plans. The mid-term goal of the project is: to develop and reach an agreement on long-term strategic action plans with clear objectives, reasonable costs and high operability; to address the priority issues and concerns about the environment in the South China Sea and its coastal areas.

The implementation content of the project includes six special subjects of mangroves, coral reefs, seagrass, wetlands, fisheries resources and land-sourced pollution control. China has participated in four special subjects of mangroves, seagrass, wetlands, and land-sourced pollution control. The participating countries of the project coordinate with each other to jointly protect the South China Sea environment, ensuring sustainable and coordinated development of society, economy and environment in the South China Sea area. The planning cycle of the project is five years, and the project was led by the State Environmental Protection Administration (SEPA) inearly March of 2002.

Regional Plan for Partnerships in East Asia Environmental Management (East Asia Project)

It is an environmental management project funded by the Global Environment Facility and organized and implemented by UNDP, mainly for the East Asian countries. The East Asia Project currently includes three phases and China has participated in all the three phases. The first phase of the project was to implement the plan of preventing pollution to the marine

On December 15th, 2006, the Haikou Partnership Agreement on the Implementation of Sustainable Development Strategy for the Seas of East Asia was signed by ministers from 11 countries at a maritime search and rescue ship in the Qiongzhou Strait, Hainan Province.

environment in East Asia (1994-1999), during which a demonstration area was established in China's Xiamen City; the second phase was to carry out the plan of establishing partnerships in East Asia environmental management (2000-2006), and address environmental management issues in hot spot waters across administrative boundaries and establish partnerships between relevant departments in environmental management. The project set up demonstration areas in Liaoning Province, Hebei Province, Shandong Province, Tianjin Municipality and Xiamen City; the third phase of the project was the implementation of East China Sea sustainable development strategies (2008-2011) in Liaoning Province, Hebei Province, Shandong Province, Tianjin Municipality and Xiamen City, as well as China's 10 model cities of integrated coastal zone management.

Action Plan for Environmental Protection, Management and Development in Northwest Pacific Ocean and its Coastal Areas (Northwest Pacific Ocean Project)

It is an integrated part of the UNEP regional seas programs. China has participated in six programs, namely: the comprehensive database and management information system program, the program of national environmental policies, regulations and strategies within the region, the program of freshwater environment monitoring and evaluation in offshore and coastal areas, the marine oil pollution preparedness and emergency response program, the marine environmental protection and public education program, the program of protection of the marine environment from land-based activities.

Yellow Sea Large Marine Ecosystem Project

It is a regional project funded by the Global Environment Facility, implemented by UNDP, and jointly executed by China and South Korea. The project, through cross-border diagnostic analysis, identified problems faced by the large marine ecosystem of the Yellow Sea, formed national and regional Yellow Sea strategic action plans, and facilitated the implementation of the action plans, so as to effectively mitigate the pressures from social and economic development in the area, advance the sustainable development and utilization of the large marine ecosystem of the Yellow Sea, and promote the social and economic development in countries bordering the Yellow Sea.

Biodiversity Management Project in South China Sea Area (South China Sea Biodiversity Project)

It is a project funded by the Global Environment Facility, implemented by UNDP, and executed SOA. The National Oceanic and Atmospheric Administration of the US has provided partial funding and technical support. The project has set up 4 demonstration areas in 5 provinces and regions in southeastern China, namely Nanji Islands National Marine Nature Reserve in Zhejiang Province, Dongshan - Nanao biological migratory corridor demonstration area across Fujian-Guangdong border, Guangxi Shankou - Weizhou Island ecological complex area and Sanya coral reef national nature reserve in Hainan Province. The project is expected to develop eco-tourism in the demonstration areas, control pollution from the land and coastal areas, restore coral reef and mangrove marine ecosystems, improve local government officials' and residents' awareness of biodiversity protection, and strengthen mechanisms for good cooperation and coordination between provinces, so as to improve the coastal areas' ability to protect the marine biodiversity.

China's Capacity for Sustainable Marine Development

The essence of sustainable marine development is coordinated development of the three pillars of marine economic, social, and environmental resources. The implementation of sustainable marine development strategies has brought huge economic, social and environmental benefits to China, especially to the coastal areas, promoting the continued improvement of China's capacity for sustainable marine development.

Enhanced Ability of Access to Main Marine Resources

With rapid development of marine science and technology, China's marine resources have strengthened their ability to support sustainable development.

Offshore Oil and Gas Resources

Over the past 30 years, China's technological innovation has boosted rapid offshore oil development, and now China is already able to explore and develop oil at the water depth of up to 3,000 meters. In 2012, the deep-water drilling platform, "Offshore Oil 981", successfully completed its first drilling, marking that CNOOC had certain capabilities for deep-water exploration and development. By the end of 2010, the cumulative crude oil reserves proved in China's offshore areas reached 4.9 billion tons of oil equivalent.

Gas Hydrates

China's sea areas have broad prospects of gas hydrate resources. At present, China has determined 11 natural gas hydrate resource prospecting areas in the continental slope of the South China Sea; the resources amount to 18.5 billion tons.

Marine Renewable Energy

Except Taiwan, the total reserves of marine renewable energy in China's offshore areas are 1.580 billion kilowatts, with a total installed capacity of 647 million kilowatts.

Marine Living Resources

China's offshore areas are rich in marine living resources. Currently, the marine industry features basically stable inshore fishing, continued development of offshore fishing resources, remarkable achievements in the mariculture industry and ascendant development of marine drug resources. The area of mariculture zones of potential development value in China is up to 1.7078 million hectares.

Sea Water Resources

While making critical progress, China's desalination research has formed an industrial development pattern with low-temperature multi-effect distillation and reverse osmosis as the two mainstream technologies. At present, about 6 billion tons of seawater is directly used in China each year, and about 600,000 tons of seawater is desalinized per day.

On March 19th, 2013, China's first 3,000 m deep water drilling BOP stack won the Innovation Award of the 13th CIPPE held in Beijing.

International Marine Resources

The international seabed areas and high seas contain a wealth of resources. Currently, the main species proved include polymetallic nodules, cobalt-rich crusts, polymetallic sulfides, gas hydrates, and rare earth, and the reserves of some minerals are dozens of times of those on land. China actively supports the mineral exploration in international seabed areas. In 2001, China obtained the priority right to explore and develop 75,000 square kilometers of polymetallic nodules. In 2010, China again obtained the exclusive right to explore and the priority right to develop 10,000 square kilometers of polymetallic sulphide mine in the southwest of the Indian Ocean.

Marine Economic Witnesses Growth in a Fast Speed

China's east coast is the most developed area of China, with rapid economic development. In 2012, 11 coastal provinces, municipalities and regions achieved over RMB 31 trillion in GDP, accounting for more than 60% of the national GDP in the same period. The sustained and healthy development of the regional economy provides a strong support for China to implement marine development and develop marine economy.

In recent years, the marine economy has showed good and fast development momentum. The total marine production grows rapidly from RMB 2.1592 trillion in 2006 to RMB 5.0087 trillion in 2012, an annual average increase of about 13%, accounting for 9.6% in the national economy. The marine economy has achieved comprehensive development and the marine economic strength has increased markedly. A complete industrial system with highlighted advantages has been formed, taking marine fisheries, marine transportation, coastal tourism, offshore oil and gas, and marine vessels as the priority, and taking marine chemicals, marine engineering and construction, marine biomedicine, and marine science and education services as important support.

The marine economic development is becoming more and more reasonable and the regional layout is gradually formed. After years of development, the marine spatial layout for the marine economic development has been shaped, mainly in the Bohai Sea, the Yangtze River Delta, and the Pearl River Delta economic zones. The economic zones develop different types of distinctive marine economies according to their own conditions. The provinces and municipalities are actively developing relevant plans and building distinctive marine economic zones. For example, Shandong has formulated and followed the Shandong Peninsula Blue Economic Zone Development Plan, while Tianjin has formulated and followed the Overall Program for Comprehensive Reform Pilot in Tianjin Binhai New Area to establish the Binhai New Area, based in the Tianjin Port, relying on the advanced manufacturing and high-tech industries, and featured by modern finance and business logistics.

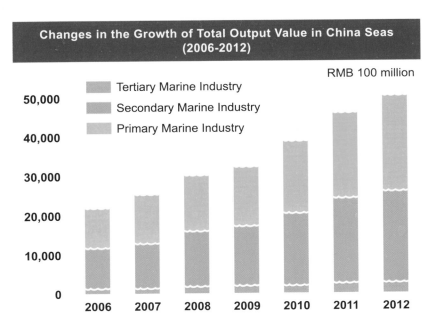

Source: State Oceanic Administration, China Marine Statistical Yearbook, 2006-2012, National Marine Economic Statistics Bulletin, 2012.

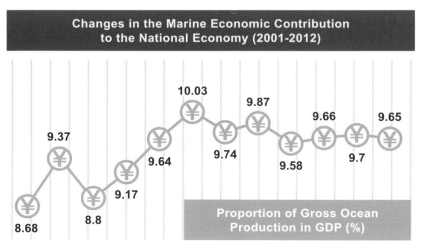

Source: State Oceanic Administration, China Marine Statistical Yearbook, 2002-2012, National Marine Economic Statistics Bulletin, 2012.

Marine Ecological Protection: from Slogan to Action

Due to the limitations of environmental awareness and of economic development, in a very long time, China maintained the situation of much said but little done in marine environmental protection. However, in the past 10 years, especially in the 21st century, China has become an actor in ecological and environmental protection, including marine environmental protection.

"Environmental Crisis" on Land

In early 2005, China's SEPA halted 30 projects under construction with a total investment of more than RMB 117.9 billion. On January 24th, SEPA announced that among all the projects called a halt to, 22 projects had been halted for rectification, but 8 illegal projects were still going their own way. Thus, SEPA issued a stern warning to enterprises including China Three Gorges Corporation, requiring them to conduct rectification within a limited time. On February 2nd, SEPA again announced that the remaining eight projects had all stopped work for rectification. SEPA Deputy Director Pan Yue said: "We will never relent in investigation of illegal projects!"

This move was called an "environmental protection storm" by Chinese and foreign media. According to a poll made by CCTV, 52% of respondents believed that the storm should have come much earlier.

Economic development is important, but a people-centered and sustainable community cannot see economic development as a "conceited" and supreme goal, but should pursue coordinated development of economy, society, resources and environmental protection, and should bear in mind that sustainable development means meeting humans' need today without compromising future generations' potentials to fulfill their needs.

This is the first time that China announced illegal projects at such a large scale. It should be said that it highlights the Chinese government's resolution to curb pollution from the source and achieve sustainable development. This resolution has also exntended its coverage from the land to the sea.

In a developing country like China, marine ecological protection has never been an easy task. However, China is using its own efforts to prove to itself and to the world its determination and achievements in marine ecological protection.

Marine Ecological Red Lines

In 2014, China held a national work conference on marine ecological protection. On the conference, some provinces and municipalities summed up their experience in drawing red lines for marine ecological protection.

In China, the drawing and adherence to the marine ecological red lines is a strategic management approach that handles the relationship between protection and development using a legal thinking, and implements both protection and development using systems. This arrangement will provide an institutional guarantee for safeguarding the marine ecosystem health and ecological security.

Shandong Province is the first province in China to promote the establishment of the marine ecological red line system.

On November 22nd, 2013, the Standing Committee of the People's Government of Shandong Province made a specialized research on the marine ecological red line system, and agreed to establish and implement the system in the jurisdiction of the Bohai Sea. On December 13th, the General Office of the People's Government of Shandong Province formally issued the Opinions on Establishing and Implementing Bohai Sea Ecological Red Line System. Together with the establishment of the Bohai Sea Ecological Red Line System in Shandong Province, a provincial joint meeting was also established for the implementation of the system.

Compromises were achieved. In May 2013, Shandong Provincial Oceanic and Fishery Department entrusted Qingdao Institute of Marine Geology under the Ministry of Land and Resources with the preparation of the Plan of Shandong Province for Drawing Bohai Sea Ecological Red Line. In June, Shandong Provincial Oceanic and Fishery Department was entrusted by the People's Government of Shandong Province to draft the Opinions of the People's Government of Shandong Province on Establishing and Implementing Bohai Sea Ecological Red Line System, and began to seek opinions from 11 sea-related departments and 4 municipal people's governments in regions around the Bohai Sea. Such work was finished until the end of September 2013.

The main contradiction between provincial governments and municipal people's governments is the conflict between protection and development. For example, some local governments have made port planning for a place, but provincial governments think the place is in a fragile and sensitive area, and hope it can be included into the ecological red line zone, thus causing a contradiction. A staff member said, in this case, Shandong Province generally followed the Technical Guide for the Drawing of Bohai Sea Ecological Red Line formulated by SOA, and both sides met each other halfway.

Cui Hongguo with the Shandong Provincial Oceanic and Fishery

Department said, "For example, some regions such as Laizhou and Longkou were previously designated as protected areas, but they wanted development. Finally, without prejudice to the principles, the two sides reached an agreement: For some regions already designated as protected areas, we can consider development according to the actual conditions; for those to be developed, we insist on including them into the red line zone."

Top-level design was adopted. In developing this red line policy, "The key point is who will lead the work when the work starts," said a staff member of SOA.

In October 2012, SOA issued a document to people's government offices of Liaoning Province, Hebei Province, Tianjin Municipality, and Shandong Province, proposing to establish the Bohai Sea ecological red line system and strengthen the environmental protection in the Bohai Sea. But only Shandong Province made quickresponse. In addition, Shandong Province is the first province in China to promote the establishment of the marine ecological red line system.

The survey made by relevant departments showed that Shandong Province's rapid implementation was related to the attention of main leaders in Shandong Province. Prior to this, China's State Council determined the objective of "ensuring the ecological safety of the Bohai Sea, reducing the total quantity of pollutants discharged into the sea, striving to improve the overall water quality in the Bohai Sea coastal waters, and working hardto achieve harmony between man and the sea". "Drawing ecological red lines", "the Bohai Sea area," and "the most stringent environmental protection policies," all these strong statements highlight the central government's resolute attitude towards the delineation of the Bohai Sea ecological red line. It was such a top-level design that strongly promoted the construction of the ecological red line system in Shandong Province.

Implementation is crucial. The drawing of the marine red line is not the final goal, and implementation of management after the red line delineation really accounts. This requires both guarantee of the total quantity, and establishment of a monitoring network or monitoring platform, and long-term mechanism for hierarchical management.

Take Shandong Province as an example. The marine ecological red line system of Shandong Province has defined 73 red line zones, of which 23 are prohibited for development and 50 are limited for development, covering a total area of 6,534.42 square kilometers. Meanwhile, Shandong Province has put forward several key objectives for the red line zones: The area of marine ecological red line zones accounts for no less than 40% of the area under the jurisdiction of Shandong Province; the retention rate of natural shorelines is no less than 40%; by 2020, the reaching-standard-rate of seawater quality within the marine ecological red line zones should be no less than 80%, the reaching-standard-rate of land-sourced pollutants discharged directly into the sea within the marine ecological red line

Tourists are visiting the national key scenic spot in the seashore area of in Shandong Peninsula, Penglaige.

zones should be 100%, and the total quantity of land-sourced pollutants discharged into the sea should reduce by 10% - 15%.

These indicators are fixed targets and are required to be achieved through management and control within the red line zone planning period. The drawing of red lines will surely sacrifice the development opportunities of some people and some regions.

Shandong will strengthen management of protected areas and protection of typical ecosystems within the red line zones, implement ecological remediation and restoration projects, carry out integrated coastal zone management, adhere to intensive and focused use of the sea and strictly control the use of the sea within the red line zones; it will also tighten up management of rivers running into the sea and outfalls, enhance control of pollutant emissions, and adjust and optimize the industrial layout; it will build and improve monitoring network and evaluation systems, strengthen environmental supervision and law enforcement within the red line zones, and enhance prevention of disasters such as red tides and emergency disposal of oil spill accidents.

These measures focus on both protection and restoration, strictly control pollutant emissions and strengthen monitoring and law enforcement so as to offer grid control over implementation of the red line system.

"It is a huge task to strengthen control of pollutant emissions," said an official with a department in Shandong Province related to marine ecological environment protection. "Environmental protection departments have done a lot to control the total quantity of pollutants. Their specific approach is to assign targets to related cities which then assign them to enterprises. So in fact, the objective isreduction of the total emissions of enterprises, instead of reduction of the total emissions of drainage areas. We want to coordinate the land with the sea and perform supervision together with the environmental protection departments. However, at present, we can only supervise some sections close to estuaries, or some estuaries and outfalls directly leading to the sea. We cannot supervise upper reaches

In June 2013, China MSA organized the MAS forces in Shandong, Hebei, Tianjin and Liaoning to carry out joint law enforcement operation in Bohai Sea so as to crack down illegal and transport in the sea and protect its safety.

which are not within the scope of our duties. Next, we hope to rely on provincial government joint meeting mechanism to coordinate with the sea-related provinces and departments in pushing forward the work; determine the implementation objectives and units responsible for the tasks— governments at various levels in regions around the Bohai Sea, assign the objectives and tasks to specific units, and promote effective implementation of the system; gradually establish ecological evaluation systems within marine ecological red line zones and highlight the conservation value of the red line zones; take the implementation conditions of the targets into account to assess government officials above the county level who are in charge of the tasks; encourage and guide enterprises and private capital investment, and improve public participation mechanisms."

Significant Progress

In recent years, China has attached great importance to marine environmental protection. Through adhering to the principle of coordination of land and sea and overall consideration of rivers and oceans, China has constantly improved laws and regulations on marine environmental protection, and strengthened prevention of marine pollution and conservation of marine biodiversity. Support systems for marine environmental protection systems continue to improve; business-oriented environment monitoring systems continue to expand, integrated remediation of land-sourced marine pollution continues to be strengthened, and the abilities to prevent and control environment damage brought by marine activities, and to respond to climate change and prevent and mitigate disasters have been quickly improved. Against the background of high-speed economic development in coastal areas, the deterioration of coastal pollution has been initially curbed, the local environmental quality has been improved, and important marine ecosystems have been effectively protected.

The meteorological station in China's Yongshu Reef

In recent years, China's marine environmental protection work has made significant progress, such as:

(1) The claim for ecological damages caused by Penglai 19-3 oil spill accident in the Baohai Sea was successfully settled. ConocoPhillips and CNOOC paid RMB 3.033 billion for ecological damages, compensation for fishery resources and restoration of the Bohai Sea environment.

(2) SOA published in print the Opinions on Establishing Bohai Sea Ecological Red Line System, designated important marine ecological function areas, ecologically sensitive areas and ecologically fragile areas as key control areas, and implemented systems for strict classified control of the areas.

(3) Marine ecological civilization demonstration zones were constructed and the strategic deployment about "ecological civilization construction" proposed on the 18th National Congress of the Communist Party of China was implemented.

China has gradually established a climate change observation network. During the "Eleventh Five-Year Plan" period (2006-2010), China completed the upgrade of more than 70 ocean observation stations and the transformation of 5 ships for monitoring the exchange flux of air-sea carbon dioxide, and regularly organized monitoring of the exchange flux of air-sea carbon dioxide in 20 sections in waters under the jurisdiction of China. Meanwhile, China actively took the advantage of polar investigation and ocean survey to conduct cruising observation, to initially form a three-dimensional ocean observation network containing land-based (island-based) stations, buoy stations, shipping platforms and satellite remote sensing, covering China's offshore areas and some major oceans.

To resist typhoons, storms and other marine disasters, China has actively built systems for disaster prevention and mitigation. During the "Eleventh Five-Year Plan" period (2006-2010), China initially established an emergency management system to deal with marine disasters covering marine departments at three levels of the state, sea areas and provinces (autonomous regions and municipalities), and the capabilities for observation, early warning and prevention of marine disasters were significantly enhanced. To grasp in time the situations of sea level rise, coastal erosion and sewater invasion, China re-approved the benchmark tide levels of 94 coastal tide gauge stations and the warning water levels of major sections in coastal areas.

Of course, China's marine environmental protection work still has a long way to go.

Breakthrough in Marine Science and Technology —A New World Record for Operational Manned Submersibles

On June 27th, 2012, China's manned deep-sea submersible, "Jiaolong", completed its deepest dive of 7,062 meters below the sea, creating a world record for operational manned submersibles; the use of Ocean Oil "981" marked a new breakthrough in China's deep-sea technological development.

China's manned deep-sea submersible "Jiaolong"

In April 2014, the Chinese Wind Energy Association released the China's Installed Wind Power Capacity in 2013. The statistics show that by the end of 2013, China had built offshore wind projects with a total installed wind energy capacity of 428.8 MW, with the installed wind energy capacity in intertidal zones being 300.5 MW and that in offshore areas being 128.1 MW. In 2013, the added installed capacity was 39 MW, representing a YoY decrease of 69%.

It should be said that China's marine science and technology has initially entered a coordinated development period. The overall strength of marine science and technology has increased; some areas have reached an international advanced level; the number of achievements, papers and patents gaining national awards has greatly increased; the conditions and environment for marine science and technology innovation have been improved remarkably; marine science and technology teams and basic infrastructure platforms have been formed to provide strong support for sustainable marine development.

In the "Twelfth Five-Year Plan" period (2011-2015), the investment in marine science and technology is increased steadily and the output capacity of marine science and technology is improved and continues to achieve new breakthroughs.

Foundation of National Ocean Council and Initial Establishment of Integrated Marine Management

In response to the initiative in the Agenda 21 adopted on the 1992 UN Conference on Environment and Development, the Chinese government promised to establish a comprehensive marine management system. Since the beginning of the 21st century, China has continually perfected the marine management system, improved the legal systems, strengthened the law enforcement management, and increased the integrated ocean management and control capabilities. China has initially established a relatively complete system of marine policies, laws and regulations. Taking the Marine Environmental Protection Law of the People's Republic of China, the Law of the People's Republic of China on the Administration of Sea Areas, the Island Protection Law of the People's Republic of China, the Fisheries Law of the People's Republic of China, and other supporting regulations as the core, the legal system has provided strong legal protection for sustainable marine development in China. The implementation of marine development, the construction of a maritime power, and the development of national strategies has provided new requirements and new opportunities for establishment and improvement of a comprehensive marine management system.

In 2013, the first session of the 12th National People's Congress decided to reorganize SOA and set up a high-level deliberation and coordination agency, National Ocean Council. Thus the comprehensive marine

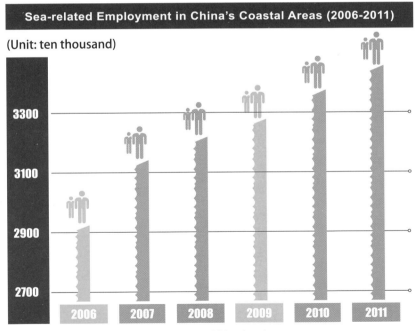

Source: SOA's China Marine Statistical Yearbook

management system was greatly improved. The establishment of a marine planning system at two levels marked that China's comprehensive marine management and integration of economy, technology, and environment has entered a new phase. Further, the issuance and implementation of marine economic development plans in the four economic zones of Guangdong, Zhejiang, Shandong, and Fujian provinces also marked the initial establishment of China's land and sea coordinating mechanism.

Marine Industries to Support Local Employment and Community Development

In 2013, the output value of major marine industries in Yantai City of China's Shandong Province reached RMB 205.41 billion, an increase of 18.9% over the previous year, accounting for 17.3% of the city's GDP, and representing an increase of 1.6% compared with the previous year.

Among the said major marine industries, emerging marine industries such as marine bio-pharmaceuticals, comprehensive utilization of seawater and marine power are growing rapidly. According to statistics, the output value of marine fisheries reached RMB 68.48 billion, an increase of 13.6% over last year; the annual growth rate of ships and marine machinery manufacturing was 45.2%; the marine biopharmaceutical industry witnessed fastest growing, an increase of 83.47% over the previous year.

Marine industries have provided important employment opportunities and income sources for people in coastal areas in China. The employed population related to sea in China increased from 29.60 million in 2006 to 34.20 million in 2011. Marine industries have kept intensifying its support for local employment.

On some islands where economic and social development is lagging behind, locals tend to rely on the sea for their livelihoods. With the popularity of mariculture and the development of tourism, small-scale capture fisheries, aquaculture and household tourism services have become the income pillars of local residents. Sea-related operation has improved the economic level of remote islands and the income level of the residents there. With the support of national policies, the infrastructure on some remote islands has been significantly improved, and problems relating to local people's living standards, such as drinking water supply, electricity supply and road construction, have basically been solved.

Sustainable Development Path of China' Seas

In China, the marine industries are at a critical development stage, moving from economic growth points to leading industries. The outline of the national "Twelfth Five-Year Plan" (2011-2015) in China takes marine economic development as a strategic focus to "develop modern industrial system and improve the core competitiveness".

The next five years isan important and strategic period withopportunities to develop China's marine economy, and also a critical time for China to realize transformation of its marine economic development mode. The sustained and rapid development of marine economy is the core task for future sustainable marine development in China.

Undoubtedly, marine economy has become the most important component of China's national economy and is an area with infinite potentials.

However, for the above reason, the following problems in China's sustainable marine development cannot be ignored: First, from the point of view of the marine development model, China attaches more attention to offshore areas but less to high seas, focuses more on resource development but less on marine eco-efficiency, and concentrates more on immediate interests but less on long-term planning; second, from the perspective of regional distribution of marine development, the industrial levels are similar (homogeneous) and the industrial structures are converging (isomorphic); there are more traditional industries but few emerging industries; there are more high energy-consuming industries but less low-carbon industries; third, since the heavy chemical industry is located in coastal areas, there are huge potential environmental risks in the sea and coastal areas. In the harbor industrial zone, there are basically steel, petrochemical, nonferrous metals, machinery, automobiles projects; large thermal power plants, nuclear power plants, oil refineries, offshore oil and gas pipeline projects and national oil reserve bases have been built and expanded along the coast, showing a centralized and scaled trend, and bringing enormous potential ecological risks of thermal pollution, nuclear leaks and oil spills to adjacent waters; fourth, climate change has become one of the most serious challenges for

In Sanya fishing port, fishermen reaped a bumper harvest.

sustainable human development. In 2012, the direct losses caused by storm surge disasters only reached RMB 12.629 billion.

Improving the capabilities for marine resource development, developing marine economy, and protecting the marine environment are the basic principle for future sustainable development of China's seas. In order to achieve sustainable development, China will take the road of coordination between land and sea, development of the sea to enrich and strengthen the country, harmony between man and the sea, cooperation for development.

Promoting the Strategy of Marine Economy Relying More on Leading Industries

Marine economy has become a new field of the national economy. Its development is at an important stage moving from the economic point of growth to the direction of a leading industry. The "Twelfth Five Year Plan" has considered the development of the marine economy as a strategic focus for the national "development of modern industrial system and the enhancement of core competitiveness". The following five years will be a significant period of strategic opportunity and an important fortified period for the transformation of marine economic development patterns. Keeping the rapid development of marine economy is the core task of Chinese future marine sustainable development. The development of China's marine economy must be in the overall strategic framework of national economic development,conform to the world's blue economic development trends, establish a modern industrial system of marine with international competitiveness, and promote the sustained and healthy development of marine economy.

To promote marine economy to a leading industry calls for theestablishment of a healthy and vibrant marine industry system. Weshould upgrade and transform traditional industries, actively deploy future industry, and improve the supporting role of science and technology innovation and management innovation to marine industry.

Breaking the Shackles of Traditional Marine Conceptions and Materalizing the Removal of the Old and Establishment of the New in Marine Industries

Traditional marine industry plays a decisive role as an important cornerstone of people's livelihood. To a certain degree, there exist somestructural problems which hinder the development in fishery, marine oil and gas, coastal tourism industry, marine transportation, shipbuilding and other traditional industries. Marine fishery and aquaculture mainly concentrate in the shallow water near the bank. Offshore fishery accounts for more than 90% of the total. The continuous decline of fishery resources, the rough mode of production and the weak ability toresist natural disasters are still the bottlenecks of development. The capacity for exploring and mining is relatively weak and the discovery rate of offshore oil is less than 20%; there exist the lack of deep sea mining equipment and the gap of the offshore drilling platform, thehigh-end industry and design capability of a variety of special shipbetween China and advanced countries. The development of coastal tourism industry booms, but the main problem it faces is the obvious homogenization and the facility lag. In the shipping industry, there is excess capacity crisis; shipbuilding industry faces overcapacity and needs technology innovation; port construction faces the project ofmarket transformation and upgrading; port-centered industry is also faced with serious environmental protection mission. To enhance and transform traditional industries is an essential task for promoting the healthy development of marine economy.

Facing fluctuations in the shipbuilding market, some far-sighted business owners have begun to make changes. When reporters of China

Offshore wind turbine of Rudong wind farm, China Longyuan Power Group Cooperation Ltd., China Guodian rotated.

Industrial and Economics News interviewed Shanghai Waigaoqiao Shipbuilding Co. Ltd., they discovered that In order to cope with the unpredictable market, Shanghai Waigaoqiao Shipbuilding Co. Ltd also takes measures to promote the company's business restructuring and upgrading of product structure. Oriented by customers' demands, it takes the cost focus strategy and moderate related diversification strategy to enter more segment of shipbuilding, marine and non vessel and gradually optimize the company's industry and product structure.

The transformation of traditional marine industry needs the reformation of the main production as well as the guidance and support of national policy. From 2014 to 2016, China will support the marine engineering equipment industry with "package" policy. Thus, nine departments jointly write the Marine Engineering Equipment Project Implementation Planincluding China NationalDevelopment and Reform Commission, Ministry of Finance, the Ministry of Industry and Information Technology. The Plan specifies safeguard measures in six aspects, and the most powerful one is fiscal measures which include deduction of part of the enterprise's R & D expenses from the taxable income, concessions on the key parts and raw materials tariff and value-added tax and the risk compensation mechanism for using the first homemade (set) homemade product.

Deployment of Future Marine Industries from a Strategic Perspective

China strengthens its support for high technologies in green mariculture and aquatic product processing, marine medicine and biological products, desalination and comprehensive utilization of seawater, ocean wind power, and ocean monitoring and information services that already have a strong technical foundation and specified application goals, as well as a significant role in promoting development of marine economy, so as to form a production scale and a new economic growth point soon, which is already a consensus and a top priority of the whole country.

Speeding up Cultivation of the Strategic Marine Emerging Industry

On October 10th, 2010, China's State Council promulgated the Decision on Accelerating the Fostering and Development of Strategic Emerging Industries, specifying that strategic emerging industries are important forces to lead future social and economic development. Developing strategic emerging industries has become a significant strategy of major countries in the world to seize the high ground of a new round of economic and technological development. As our country is now in the critical stage of comprehensively building a well-off society, we must, as required by the scientific outlook on development, seize the opportunity, clarify the orientation, stress the key points, and accelerate the fostering and development of strategic emerging industries.

The Decision put forward the idea of accelerating the development of the marine biological technology along with product R & D and industrialization, it advocates for the development of marine resources and marine engineering equipment. The Decision is an important guiding policy for Chinese future industrial development, and it also means a lot tothe deployment of marineindustry development.

There are seven strategic emerging industries that the Decision advocates that we should focus on: Energy saving and environmental protection, new generation information technology, biotechnology, high-end equipment manufacturing, new energy, new materials and new energy

Partly constructed Zhoushan Marine Science Technology and Creative R & D Area

automobile industry. Most of them are connected to marine. The cultivation and development of strategic emerging industries of the ocean is an important way to promote the establishment of a complete industrial system of marine and to cultivate new growth point of national economy.

Marine Drugs and New Marine Biological Products

Developing and strengthening the marine life breeding and healthy culture, marine medicine and functional food industry, high-end ship and marine engineering equipment manufacturing, modern marine services, and marine renewable energy industry, and accelerating the industrialization process of seawater desalination and utilization so as to stimulate green development of marine economy will be the focus of China's marine industry development.

90% of the species on Earth are living in the ocean where there are more than 100 million species and fewer than 20 million species are already identified and named. For the future of humanity, research and development of marine drugs and new types of marine biological products is the only way out.

Pfizer and Bristol-Myers Squibb in the U.S., Smithkline Beecham in the UK, and Roche in Switzerland are the world's leading companies in marine drugs and biological R&D and production. While creating significant economic value, they have also contributed to human health. Their success in all aspects has provided a valuable reference for economic and social development of emerging countries.

In late December of 2012, China's State Council issued the Bio-Industry Development Plan, pointing out the need to "strengthen development and utilization of living marine resources", accelerate development of unique marine biological resources, significantly enhance the ability to develop new varieties of marine aquaculture, and make greater efforts to promote application of new products; accelerate development and application of marine bioactive substances, and develop new marine products such as industrial enzymes, medical functional materials, bioseparation materials, green agricultural biological agents, and innovative drugs.

To combine production, learning and research is an important way to develop the industry of new marine biological products. In March, 2014, China's first marine microbial preparation industry development platformwent under construction in Zhaoan County of Fujian Province JinduMarine Biological Industry Park. The National Third Marine Research Institution was responsible for the construction of it with an investment of 40 million in three platforms: enzymes and enzyme by R & D, research and development of live bacteria and equipment sharingin marine microbial industry. During the construction, the park will take the research and development of microbial products by sea water as a breakthrough point and work together with the enterprise to form a chain of organic combination. It will use the existing platform to establish a set of marine microorganisms producing technology, as well as technical and qualitystandards. The project is expected to exceed the a series of key common technical problemsin the Chinese industry of utilization of marine biological resources and promote the development of marine microbial products, marine oligosaccharides and other emerginghigh tech industries.

Flourishing Marine High-tech Industries

In April 2014, China's NDRC and SOA decided to launch national marine high-tech industry base pilot in China's eight cities of Guangzhou, Zhanjiang, Xiamen, Zhoushan, Qingdao, Yantai, Weihai, and Tianjin, with the purpose of promoting high-end and cluster development of high-tech industries through pilot, facilitating optimization and upgrading of regional industrial structure, strengthening technology innovation in high-tech industries, and expanding the scale of high-tech industries.

These eight cities focus on different aspects while developing the marine high-tech industries:

China's municipality of Tianjin will focus on development of marine high-end equipment manufacturing, seawater utilization, exploration and development of deep-sea strategic resources and marine high-tech services, marine medicine and biological products, construct 5 key marine laboratories at provincial and ministerial levels or above, several marine technology R&D centers or instruments and equipment testing centers, and foster a number of leading enterprises.

Qingdao City of Shandong Province will focus on development of mariculture breeding and healthy culture, marine medicine and biological products, marine high-end equipment manufacturing, marine renewable energy, exploration and development of deep-sea strategic resources and marine high-tech services, and will basically construct a regional marine emerging industry development center and a marine science and technology and education center of the world's advanced level within a few years;

On November 25th, 2013, two semi-submersible crane platforms China independently designed and built with a complete intellectual property rights was delivered.

On December 22th, 2013, China's first floating marine test platform, the "Huajiangchi" scientific research ship hosted by Zhejiang University was accepted by the expert group and delivered to use.

Weihai City of the same province will focus on development of the two industries of marine life breeding and healthy culture, and marine medicine and biological products, and will make breakthroughs in key technologies in recent years to form a marine high-tech industry system; Yantai City of the same province will focus on development of the marine life breeding and healthy culture industry, the marine high-end equipment manufacturing industry and marine high-tech services, to form functional areas with the Yantai eastern marine economic area as the core, and form an industrial development pattern characterized by "one leading core, several supportive bases, unique parks, and joint development" supported by a few special industrial bases and unique parks.

Zhoushan City of Zhejiang Province will vigorously develop the marine high-end equipment manufacturing, marine life breeding and healthy culture and other key industries, and promote formation of a spatial

Volunteers were preparing for the Dafeng Port Marine Biological Expo, Yancheng, Jiangsu Province. It's the first comprehensive expo about marine biological industries, many compaies and scientific institutions attended the expo, including those from marine biological biomedicine, marine food, marine ecological breeding and cultivation marine bio-energy and other areas.

layout with Zhoushan Marine Science City as the core, and the northern marine scientific and technological achievement transformation belt and the southern marine science and technology innovation belt as the main body.

Xiamen City of Fujian Province will focus on development of the marine pharmaceutical and biological products industry, marine life breeding and healthy culture industry, marine high-end equipment manufacturing industry, marine high-tech service industry, to form a spatial layout extending southward to Zhangzhou and northward to Quanzhou with Xiamen Haicang Biomedical Park as the core base, and to foster one or two marine high-tech enterprises with sales revenue in excess of RMB 2 billion and several marine high-tech enterprises with sales revenue of over RMB 1 billion.

Guangzhou City of Guangdong Province will focus on development of marine high-end equipment manufacturing, marine pharmaceutical and biological products, marine renewable energy and other industries;

Zhanjiang City of the same province will focus on development of marine life breeding and healthy culture industry, and build a number of standardized aquatic fingerlings production and demonstration areas and 2 marine bio-breeding industry parks.

Some of the pilot cities mentioned above rely on Key Laboratory, some are based onexisting marine industry parks, some focus on the development of new energy industry, and others on ecological breeding. We may say that these eight cities are focused high technology development and deployment of our country according to their ability in scientific research of marine and their foundation of marine industry and resource endowments, and each one has its advantages. The aim is to boom the development of marine high-tech industry.

Scientific and Rational Development of Marine Resources

Scientific and rational exploitation and utilization of marine resources and vigorous improvement of the ability to develop them are inevitable requirements to achieve sustainable marine development.China must adapt to the new trends of international marine development and build scientific marine resources management and development systems. China should focus on improving the ability to develop marine resources, promoting green development, and improving international competitiveness of its marine economy; it should focus on optimizing the spatial pattern of ocean development, and coordinating land and sea resources configuration, economic layout, environmental management and disaster prevention; in addition, China should coordinate development intensity and utilization timing, and arrange offshore development and high sea space expansion.

As a sea power and economic power, China aims at the major problem of natural science in ocean areas, speeds up the basic research and strives to make breakthroughs in knowledge innovation to contribute to the progress of human civilization. These are not only the must of Chinese society to develop its own economy, but also human's bound duty. To put this must and responsibilityinto the action means to enhance the marine foundation, to look forward, to improve key technology research and ability to transform with technology innovation as forerunner. Strengthen the study of deep-sea biology gene and improve the research level of ocean exploration technology. Strengthen technical research in marine disaster prevention and improve the accuracy and skill of marine forecast. Construct the comprehensive survey marine security mechanism of a normalization and constantly enrich and update the marine data and information.

Ranking Zoning of Wind Energy Resource: Improve the Level of Development and Utilization of Maritime Wind Energy

To improve to the ability of developing and utilizing ocean, China enhanced the combination of producing, learning and researching in exploration, overcame the technical bottlenecks met in upgrading traditional

On April 13th, 2014, the desalination equipment with China's independent intellectual property rights and single largest capacity operated in Cangzhou City, Hebei Province.

marine industry and cultivating emerging marine industry. Surrounded the key technologies, such as offshore petroleum exploration, pelagic fishery, seawater aquiculture, seawater comprehensive utilization, salt-tolerant crops cultivation and so on of marine industry to enhance science and technology in exploration, and improve its contribution rate in marine development and utilization.

For a long time, the rapid development of coastal areas in China was restricted by energy bottleneck. Therefore, it's of great economic values to take advantage of wind energy resources in coastal areas and develop offshore wind resources. Because of the obvious regional difference of offshore wind resources' distribution, the basic rule of developing wind power in large scale is assessing and planning resources first. Only make scheme of wind power development and associated power network construction based oncareful reconnaissance of wind energy resources, can we realize ordered development and utilization of wind energy resources.

In March 2014, China achieved breakthrough on the aspect of assessment of offshore wind energy resources. Chinese young scientists Zhen Chongwei and Pan Jing broke through bottleneck and first achieved "ranking division of wind energy resources" in the range of global oceans.

This achievement was acknowledged by the international prestigious energy journal Renewable and Sustainable Energy Reviews. It shows that China is among the best on assessment of offshore wind energy resources. This study considered comprehensively in directions of seasonal characteristics of wind energy density, rate of energy level, frequency of effective wind speed, stability of wind energy density and so on, first achieved systematic assessment of global oceans' wind energy resources, which can provide scientific basis for offshore wind power, and accelerate its steps of development from near-shore areas to deep sea and remote islands.

The Sea Horse: Marching forwards to both Deep Sea and High Sea

It is mentioned in the marine sciences and technology development

On June 8th, 2014, volunteers were carrying out the clean ocean action under the theme "Blue Ocean, Eco Action" in the Haiyu Square, Longfengtou, Pingtandao Island, Fujiain Province.

of national 12th Five-Year Plan that "ocean exploitation steps into three-dimensional exploitation stage. When developing further and utilizing traditional marine resources, we need to keep exploring and developing deep sea and far offshore to get strategic new resources and energy, and extend the zone of ocean economy development." Deep sea and far offshore are not only important and potential fields of future marine resource exploitation, but also important safeguards to promote ocean economy development and obtain benefit from ocean. Therefore, it is necessary to develop deep ocean and far offshore engineering and marine industry.

From February 20th to April 22nd in 2014, the first 4,500m level deep-ocean robot unmanned underwater vehicle operating system developed independently by our country—"Sea Horse" ROV travelled by "Ocean 6" integrated science research vessel to practice marine experiment in three segments in the South China Seaand passed acceptance. In the family of underwater robots, High Occupancy Vehicle (HOV), Remote Operated Vehicle (ROV), and Autonomous Underwater Vehicle (AUV) are three most important underwater vehicles. The previous "Jiaolong" is HOV. The maximum size of ROV is less than 0.4m, its weight is less than 40kg, but its operating radius can reach 100m, which is the important device of exploration in deep oceans. The success of marine experiment of "Sea Horse" symbolized that China has mastered every key technique of unmanned underwater vehicle in great depth, and achieved substantial progress onlocalization of key technologies. It's another symbolic achievement our country achieved after "Jiaolong" HOV in deep-ocean high technology, fulfilled "the zero breakthrough" on independent development of unmanned robot underwater vehicle in great depth.

However, the development and experimental application of "Flood Dragon" and "Hippocampus" cannot change the reality that China's marine investigation equipment is laging behind. There is a large gap in deep sea survey and development ability between China and the most advanced countries. At present, expedition vessels in China for deep ocean investigationare no more than 10 in total, the majority of which is old one. China rely largely on import from the survey system used to high-edged prospecting instruments.

Therefore, it becomes more urgent to hoist capacity of exploiting deep sea and far offshore resources.conducting a survey of deep ocean and far offshore and polar exploration, producing key techniques of deep ocean resource exploration and research on series of submarine technical equipment and industriali-zation, establishing deep ocean and far offshore environmental monitoring network, sea orbiter, polar station, large seagoing research vessels and polar-exploration craft platform, etc are tasks of developing ocean economy for China in the medium to long-term. Only with the innovative development of high technology of deep ocean which has a induced effect of radiation to the leapfrog development of relative marine technology, can we really push the development of China's marine industry and the transition of marine economy.

Contrusting the Ecosystem-based Comprehensive Marine Management

Management based on ecosystem is considered to be an effective way to actively address transboundary marine development and utilization problems and protect and maintain the marine and coastal ecosystems and their functions. Through the construction of Marine Reserves, the implementation of marine functional zoning and the establishment of the high-level marine coordination agencies—National Marine Committee and other measures, the Chinese government make important progress in the establishment of the integrated management based on marine ecosystem.

The oceans do not exist alone. To a great extent, they depend on the land. So The marine resources and environment degradation, marine habitat destruction, functional conflicts of marine and coastal zone using, the low retention rate of natural coastline still exists, which influences the sustainable development of Chinese marine. Therefore, the key to China's sustainable marine development is sticking to land and sea coordination, implementing ecosystem-based marine management, forming a management and control structure characterized by the use of administrative, legal, and economic means, combination of the central and local governments, government dominance and community involvement, and enhancing the comprehensive capabilities for marine management and control.

To further promote the comprehensive marine management based on ecological system, it is necessary to emphasize the leading role of planning and zoning, to complete the legal marine systemand establish the institutional and legal basis for the comprehensive marine management.

Model City: Strenghtening the Leading Role in Planning and Zoning

For the establishment of comprehensive marine management based on ecological system, we should insist on coordination of land and sea, improving national and provincial marine functional zoning system, well conducting the planning of marine development , marine economy, island protection jointing with other planning,division and cooperation. We should insist on strengthening the organization, implementation and supervision, inspection of marine planning and zoning, improving the development and utilization of marine macro management, optimizing space layout of marine development and making the planning, zoning regulation in economic and social development in the coastal areas into full play.

Xiamen is a famous coastal city in Fujian province, China, famous for coastal tourism industry, bio-pharmaceutical industry and modern service industry. since 1990s, Xiamen has introduced the integrated coastal zone management mechanism and has successfully controlled the deterioration of the marine environment, which has provided a better recreation environment for residents and visitors. Xiamen walks in the front of China in the aspect of the protection of the marine environment and promulgates marine environment protection law and planning such as "Provisions Xiamen Marine Environmental Protection" "Xiamen Marine Environmental Protection Plan" which lays the foundation for the system of protection of the marine and coastal zone. In recent years, Xiamen is launching new initiatives in the protection of the marine environment constantly.

On April 3rd, 2014, Xiamen officially launched "Beautiful

On September 13th, 2013, the Key Cooperation Projects Signing Ceremony of the 8th Cross-Strait Fisheries (Fuzhou) Expo was held in Fuzhou Strait International Conference and Exhibition Center.

Construction of Marine Special Action Plan (Draft) in Xiamen" and established the completion of model city of beautiful ocean in China as development objections. Xiamen possesses favorable marine environment. A comprehensive rectification of West and East Sea in 2002 and 2006 reduced the sea farming scale. The sea ports, tourism, ecological main functional areas accounts for about 80%, which greatly promotes the development of ports, coastal tourism. In addition, Xiamen vigorously promotes the marine bio pharmaceutical, yacht cruise and seawater utilization of emerging marine industry which makes the marine economic value increase from 9.46 billion yuan in 2003 into 31.85 billion yuan in 2012 accounting for 11.3% of GDP, with an average annual increase of 12%. At present, economy added-value of Xiamen city created by sea reaches 81 million yuan per square kilometer, 20 times higher of the average provincial level , over 40 times higher of the average national level.

Pushing forward the protection of the marine environment, promoting the development of the marine science and technology industry to make the seaside city rejuvenate. "Beautiful Ocean Construction of Marine Special Action Plan (Draft) in Xiamen" represents the desire for a further development of the marine industry in Xiamen Cityand the goal to achieve the development of marine economy, to beautify marine ecological environment, to upgrade marine science and technology innovation, to rich marine culture and marine management forcefully and effectively, to realize the construction of beautiful ocean and to make Xiamen a model city for the construction of China's beautiful ocean.

Xiamen is one of members of coastal city in China. With the leading and exemplary of Xiamen, China will have more coastal cities to extort strenghtens of the whole society, unify people of all walks of life and take a down-earth attitude, protect the marine environment, promote marine developmentand build the beautiful sea.

Systematic Construction in Sea Use: Improvement of Marine Laws and Regulations

Marine Environment Protection Law, which begins to take effect from March 1st, 1983, marks the protection work of Chinese marine environment has entered the law system track. With the development of enterprise of marine environment protection, the Marine Environment Protection Law is gradually sane and has formed a system, which is on the basis of Constitution, based on Environmental Law, on the body of specialized laws such as Marine Environment Protection Law, Protection Law of Wild Animals, Fishing Laws and so on, supplemented by marine environment protection executive regulations, local regulations and laws, coordinated with International Convention.

Of course, China's marine environment legal system construction hasn't been done and should continue to perfect marine laws and regulations system, on the basis of existing marine laws and regulations. For example, coming on Marine Basic Laws as soon as possible; continuing pushing

On June 8th, 2014, Activities concerning World Oceans Day and the National Marine Awareness Day held in Shanghai

forward Marine Environment Protection Law as well as revising of its auxiliary regulations; boosting the legislature procedure of the Bohai Sea Regional Management Law and the Integral Management of Coastal Area. On the aspect of implementing the launched laws and regulations, China still needs to more forward the effective action.

China Ocean News on March 20th 2014 published China's management memorabilia of costal areas' usage in 2013, among them many are about system construction of coastal areas' usage. For example, State Oceanic Administration and costal provinces and cities have already set many regulations of costal areas' usage in January, February, March in 2013, including on February 6th, State Oceanic Administration printed and distributed Main Point of Waters Regulation in 2013, on February 28th, Zhe Jiang Province came on management regulations such as Temporary Conducting Method of Application and Examination Management of Costal Areas' Usage, on March 1st, Costal Areas' Usage Regulations in Zhe Jiang Province, on March 14th, marine and fishing administration of Han Nan Province printed and distributed Application and Examination Procedure of Costal Areas' Usage and Application and Examining Procedure of Annual Plan Index of Encircling the Sea to Make Land, and so on.

Regulating costal areas' usage is the powerful tool of relieving the contradiction between resource exploitation and environmental protection. The system construction of water usage area also reflects the determination of China's proposal of exploiting marine resources and maintaining the benefit of marine ecology environment.

Construction of Marine Ecological Civilization "Blue Ribbon"

Beauty of China cannot be separated from beauty of its ocean. Constructing marine ecological civilization and tackling climate change is an important part of China's construction of ecological civilization and also a due connotation of sustainable marine development.

On June 1st, 2007, Blue Ribbon Ocean Conservation Society was founded in Sanya City of Hainan Province. A person in charge of the Society explained, "Blue Ribbon" represents gratitude, encouragement, care and love. As a private nonprofit marine conservation organization, at the beginning, the society only had more than 40 enterprise members including Nanshan Group, Tianya-Haijiao, Sheraton Hotels & Resorts, Industrial and Commercial Bank of China, China Netcom (Hainan), China Mobile (Sanya), Sanya Luneng, and Yalong Bay. Up to now, the Blue Ribbon Marine Conservation Society has 61 member units and 5 donation units. The society has established "Blue Ribbon Volunteer Services" in several universities in Hainan, Guangdong, and Shanghai, forming a team of more than 10,000 volunteers. They have organized 300 various marine conservation campaigns, issued 200,000 brochures and 300,000 marine environmental protection wristbands, and made publicity of marine protection to over 10 million audiences With participation of nearly a million volunteers.

In order to better carry out marine conservation activities, Lenovo Venture Philanthropy Foundation offered funds for the "Blue Ribbon" for a variety of marine conservation activities. In addition, Yihe Real Estate and the Chinese Academy of Social Sciences also donated funds to the "Blue Ribbon" to carry out various activities.

On June 8th, 2010, the World Oceans Day, the People's government in Sanya City, Hainan Province, carried out the "China's Blue Ribbon Ocean Conservation" activity in Dadonghai Square.

In 2013, Zhang Jinghua from the Sanya University organized and became the person-in-charge of the "Sanya River Protection" project. With the help of the Blue Ribbon Ocean Conservation Society, Zhang Jinghua gathered the four universities and all public interest groups in Sanya City to carry out the six-month "Sanya River Protection" activity, including investigation, cleaning, publicizing, online and offline coordination, and further extension from "river" to "sea", attracting lots of volunteers to carry out comprehensive publicity campaigns in schools, communities and surrounding shops.

Through field investigation and recording, the volunteers figured out the current conditions of the outfalls, pollution sources, and garbage distribution around the Sanya River, and prepared an investigation report, providing reference data for relevant government departments to carry out their work.

From cleaning beaches to coordinating experts' research, from popularizing science on marine environment and resources protection to organizing training seminars, and from protecting the Sanya River to calling for more volunteers to join the team of the marine environmental protection, Zhang Jinghua was busy with the cause of marine environmental protection.

In Meilian Village, Yacheng Town, Sanya City, there are several estuaries. Since there were only two garbage stations on the two ends of the village, some villagers threw garbage into the estuaries for convenience. Over time, the estuaries became "mountains of garbage". After the discovery of this case, Zhang Jinghua and other volunteers contacted the local village committee and sanitation bureau, borrowed garbage trucks and shovels, and removed the hazards that affected the marine environment against the scorching sun and pungent taste. In order to avoid subsequent formation of "mountains of garbage", Zhang Jinghua carried out the ocean protection and visit campaign and told the local people the hazards of garbage dumping so as to raise the awareness of local residents to participate in environmental protection.

To address the issue of staff turnover in the public welfare organization, Zhang Jinghua adopted the theory of "Applied Psychology" she had learnt, by organizing volunteer training and development activities to remove volunteers' self-doubt and waver,and strengthen team cohesion and solidarity on the one hand.On the other, she improved the organization of the team, by setting up "Volunteer Home" and created "Family Culture" allowing volunteers with outstanding abilities to manage and organize the established and new volunteers in volunteer activities, andoffer opportunities and platforms to those who want to participate in volunteer activities. These measures have not only reduced the loss of talents, but also absorbed more new volunteers.

Although China has only a few marine environmental protection organizations like "Blue Ribbon" and their influence is not so big, with raising of awareness of marine environmental protection and marine ecological civilization construction among both governmental and non-governmental organizations, we can expect that "Blue Ribbon" will soon be found everywhere along the Chinese coastline.

Assessment of Leading Cadres' Responsibility for Marine Environmental Protection

If we say that marine environmental protection has been strengthened among Chinese non-governmental organizations, then strict systems for assessing marine environmental protection are mushrooming in government departments at various levels in China.

How well is the marine environment protected in China's maritime province Fujian? Assessment items have been set up to make a judgment with scoring. This initiative of Fujian Province has aroused widespread concern in society.

Five assessment items were set up with a full score of 120 points. The Office of the Leading Group for Marine Development and Management in Fujian Province decided to carry out responsibility and objective assessment scoring in mid-April of 2014 concerning the 2013 marine environmental protection work conducted by people's governments at regional and municipal levels in coastal areas and the Management Committee of Pingtan Comprehensive Experimental Zone. The responsibility and objective assessment scores of regions and cities in the annual coastal marine environmental protection would be included by weights in 2013 total scores for the responsibility and objective assessment of the regions and cities.

Among the five assessment items, marine environment quality accounts for 20 points and this assessment item includes the quality of seawater in coastal areas within the jurisdiction, whether the seawater quality is in line with requirements of the ocean functional areas, and the quality of seawater in key estuaries; marine pollution control accounts for 35 points and the assessment content includes implementation of local marine environmental protection plans, marine environment (including emergency) monitoring, tail water discharge compliance in coastal sewage treatment plants, sea-related project EIA report and environmental protection measures, aquaculture planning and environment remediation, vessel pollution prevention in major ports, and floating garbage governance in key areas; marine ecological protection accounts for 30 points and

Cleaning up the floating oil and white trash to protect marine environment

the assessment content includes management of various protected areas, comprehensive remediation and restoration of marine environment, sea sand resources protection and management, construction of marine ecological civilization demonstration zones, and sea-related project marine ecological damage compensation pilot; marine environmental monitoring capability accounts for 15 points and the assessment content includes annual marine environmental management system and funding mechanism, establishment of land and sea environmental protection mechanism, and marine environment monitoring and enforcement capacity building at the county level; regional prominent marine environment issue remediation accounts for 20 points and the assessment content includes two of the regional prominent marine environmental issue remediation goals in the annual marine environmental responsibility goals for regions and cities in coastal areas.

The above five assessment items are assessed individually or jointly by Fujian Provincial Department of Ocean and Fisheries, Fujian Provincial Department of Environmental Protection, Housing and Urban-Rural Development of Fujian, Fujian Provincial Department of Forestry, Fujian Maritime Safety Administration, and Fujian Provincial Oceanic and Fishery Law Enforcement Team.

Conclusion

The ocean is so beautiful, defying all comparison.

The beauty of the ocean not only lies in its aesthetic value shown in aquariums and polar museums, and its wealth value as the treasure trove of resources for human survival at present and in the future, but also rests with its spiritual value of being an inherent companionship with humans, and its value beyond time and space as a sign of the existence of Earth as a planet in the universe.

With development of human beings and advance of the world, China, as a developing country, shall consciously and must look upon oceans and their development from an always new perspective and with an always new attitude while making leapfrog economic and social development.

This consciousness and must may not be formed within 108 minutes, but it has been rooted and is now growing in the hearts of more than 1.3 billion Chinese people, and is producing more significant effects to make the world and the world's oceans more beautiful.

Schedule 1

National Marine Nature Reserves of China

No.	Name of Protected Area	Administrative Area	Area (Ha.)	Main Protection Target	Type	Time of Initial Construction
1	Hepu Yingpangang - Yingluogang Dugong	Beihai City	35000	Dugong and marine ecosystems	Wild animals	1986-4-27
2	Shankou Mangrove Forest Ecosystem	Fangcheng District of Fangchenggang City	8000	Mangrove ecosystem	Ocean coast	1990-9-30
3	Beilunhekou	8000	3000	Mangrove ecosystem	Ocean coast	1985-1-1
4	Jiuduansha Wetland	Pudong New Area	42020	Estuary sandbar geomorphology and birds	Inland wetlands	2000-3-1
5	Chongming Dongtan Birds	Chongming County	24155	Migratory birds, Chinese sturgeon	Wild animal	1998-11-1
6	Huangjin Hai'an	Changli County	30000	Beach and coastal ecosystems	Ocean coast	1990-9-30
7	Ancient Coast and Wetland	Ninghe County, Dagang District, Jinnan District	99000	Shell banks, submarine forests ruins on oyster beaches, coastal wetlands	Ocean coast	1984-12-1

8	Dalian Spotted Seal	Dalian City	909000	Spotted seals and their habitats	Wild animal	1992-9-1
9	Shedao-Laotieshan	Lüshunkou District of Dalian City	14595	Viper, migratory birds and Shedao special ecosystem	Wild animal	1980-8-6
10	Chengshantou Coastal Landforms	Jinzhou District of Dalian City	1350	Geological relics, fossils and coastal karst	Geological heritage	1989-4-1
11	Yalujiangkou Littoral Wetland	Donggang City	108057	Coastal tidal wetlands and migratory waterfowls	Ocean coast	1987-7-1
12	Shuangtaihekou	Xinglongtai District of Panjin City	80000	Rare waterfowl and wetland ecosystems	Wild animal	1987-1-1
13	Binzhou Shell-dyke Island and Wetland	Binzhou City	80480	Seashell Islands, wetlands, rare birds, marine life	Ocean coast	1998-10-1
14	Huanghe Sanjiaozhou	Dongying City	153000	Native wetland ecosystem and rare birds	Ocean coast	1990-12-27
15	Changdao	Changdao County	5300	Habitats for eagles, falcons and migratory birds	Wild animals	1982-1-1
16	Rongcheng Whooper Swan	Rongcheng City	1675	Whooper swans and their habitats	Wild animals	1992-5-30
17	Xiamen Valuable and Rare Marine Species	Xiamen City	33088	Chinese white dolphins, egrets, amphioxus	Wild animals	1995-1-1

	Name	Location	Area	Main protected objects	Type	Establishment date
18	Shenhuwan Submarine Paleoforest Relic	Jinjiang City	3400	Submarine forests ruins, oyster beach rocks and geological features	Paleontological remains	1991-1-1
19	Zhangjiangkou Mangrove Forest	Yunxiao County	2360	Wetland mangrove ecosystems	Ocean coast	1992-7-1
20	Dongzhaigang	Meilan District of Haikou City	3337	Mangrove ecosystem	Ocean coast	1980-4-9
21	Sanya Coral Reef	Sanya City	4000	Coral reef and its ecosystem	Ocean coast	1990-9-30
22	Tongguling	Wenchang City	4400	Coral reefs, tropical monsoon forest, wildlife	Ocean coast	1983-5-24
23	Dazhoudao Marine Ecosystem	Wanning City	7000	Swiftlets and their habitats, marine ecosystems	Ocean coast	1987-8-1
24	Dafeng Père David's Deer	Dafeng City	2667	Elks and their habitats	Wild animals	1986-2-8
25	Yancheng Littoral Mudflats and Valuable Fowls	Dafeng, Binhai, Dongtai, Sheyang	453000	Red-crowned cranes and other rare birds, wetland ecosystems	Wild animals	1984-1-1
26	Leizhou Valuable and Rare Marine Organisms	Zhanjiang City	46864	Leizhou Bay marine ecosystem	Ocean coast	1983-1-1

27	Xuwen Coral Reef	Xuwen County	14378	Coral reef ecosystems	Ocean coast	1999-8-1
28	Huidong Gangkou Sea Turtle	Huidong County	800	Spawning and breeding ground for sea turtles	Wild animals	1986-1-1
29	Neilingdingdao - Futian	Futian District, Shenzhen City	815	Mangroves, macaques, birds	Ocean coast	1984-4-9
30	Zhujiangkou Chinese White Dolphin	Zhuhai City	46000	Chinese white dolphins and their habitats	Wild animals	1999-1-1
31	Nanpeng Liedao	Nanao County	35679	Marine ecosystems and marine animals	Ocean coast	1991-1-1
32	Zhanjiang Mangrove Forest	Zhanjiang City	20279	Mangrove ecosystem	Ocean coast	1990-1-8
33	Jiushan Liedao\	Xiangshan County	48478	Pseudosciaena crocea, birds and reef ecosystems	Ocean coast	2003-4-18
34	Nanji Liedao Marine	Pingyang County	19600	Shellfishes and algae and their habitats	Ocean coast	1986-1-1
35	Minjiang Estuarine Wetland	Changle City	3219	Marine and coastal ecosystems	Ocean coast, wild animals	2001

Schedule 2

National Special Marine Protected Areas of China

No.	Name	Area (Ha.)
1	Liyashan National Marine Park, Haimen, Jiangsu	1222.9
2	Ximendao National Special Marine Protected Area, Yueqing, Zhejiang	3080
3	Ma'an Liedao National Special Marine Protected Area, Shengsi, Zhejiang	54900
4	Zhongjieshan Liedao National Special Marine Eco-Protected Area, Putuo, Zhejiang	20290
5	Yushan Liedao National Special Marine Eco-Protected Area, Zhejiang	5700
6	Changyi National Special Marine Eco-Protected Area, Shandong	2929.28
7	Huanghekou Ecosystem National Special Marine Protected Area, Dongying, Shandong	92600
8	Lijin Demersal Fishes Ecosystem National Special Marine Protected Area, Dongying, Shandong	9404
9	Hekou Shallow-water Shellfishes Ecosystem National Special Marine Protected Area, Dongying, Shandong	39623
10	Laizhouwan Jackknife, Razor and Short Razor Clams Ecosystem National Special Marine Protected Area, Dongying, Shandong	21024
11	Guangrao Sandworms Ecosystem National Special Marine Protected Area, Dongying, Shandong	8282

12	Wendeng Ocean Ecosystem National Special Marine Protected Area, Shandong	518.77
13	Huangshuihekou Ocean Ecosystem National Special Marine Protected Area, Longkou, Shandong	2168.89
14	Zhifudao Island Group National Special Marine Protected Area, Yantai, Shandong	769.72
15	Liugongdao Ocean Ecosystem National Special Marine Protected Area, Weihai, Shandong	1187.79
16	Tadaowan Ocean Ecosystem National Special Marine Protected Area, Rushan, Shandong	1097.15
17	Muping Sandy Coast National Special Marine Protected Area, Yantai, Shandong	1465.2
18	Wulonghekou Littoral Wetland National Special Marine Protected Area, Laiyang, Shandong	1219.1
19	Wanmi Haitan Ocean Resources National Special Marine Protected Area, Haiyang, Shandong	1513.47
20	Xiaoshidao National Special Marine Protected Area, Weihai, Shandong	3069
21	Dabijiashan National Special Marine Protected Area, Jinzhou, Liaoning	3240
22	Dashentang Oyster Reef National Special Marine Protected Area, Tianjin	3400
23	Laizhou Qiantan Ocean Ecosystem National Special Marine Protected Area, Shandong	6780.1
24	Dengzhou Qiantan National Special Marine Protected Area, Penglai, ShandongSpecial Marine Protected Area, Shandong	1871.42

Schedule 3

National Marine Parks

No.	National Marine Park	Area (Ha.)	Protection Target
1	Liyashan National Marine Park, Haimen, Jiangsu	1545.91	Coastal geological landscapes
2	Haizhouwan National Marine Park, Lianyungang, Jiangsu	51455	Unique sea landscapes, special bedrock reefs, marine natural heritage resources
3	Xiaoyangkou National Marine Park, Jiangsu	4700.29	Wetland landscapes
4	Yushan Liedao National Special Marine Eco-Protected Area (National Marine Park), Zhejiang	5700	Marine fishery resources, island landscapes, fishing village culture
5	Dongtou National Marine Park, Zhejiang	31104.09	Island life, fishermen life, island landscapes
6	Xiamen National Marine Park, Fujian	2487	Coastal landscapes
7	Fuyao Liedao National Marine Park, Fujian	6783	Island landscapes
8	Changle National Marine Park, Fujian	2444	Historical and cultural sites
9	Meizhoudao National Marine Park, Fujian	6911	Mazu culture, island landscapes
10	Chengzhoudao National Marine Park, Fujian	225.2	Marine fishery resources

11	Liugongdao National Marine Park, Shandong	3828	Island landscapes
12	Rizhao National Marine Park, Shandong	27327	Coastal landscapes, historic sites
13	Darushan National Marine Park, Shandong	4838.68	Coastal sceneries
14	Changdao National Marine Park, Shandong	1126.47	Coastal geomorphology, spotted seals
15	Hailingdao National Marine Park, Guangdong	1927.26	Bay view, historical and cultural sites
16	Techengdao National Mainre Park, Guangdong	1893.20	Coastal biome landscapes, geological heritage
17	Wushi National Marine Park, Leizhou, Guangdong	1671.28	Fishing folk culture, seaside landscapes
18	Maoweihai National Marine Park, Qinzhou, GUangxi	3482.70	Mangroves and salt marshes and other marine ecosystems
19	Weizhoudao Coral Reef National Marine Park, Guangxi	2512.92	Island scenery, coral reefs